Gender, Water and Development

Cross-Cultural Perspectives on Women

General Editors: Shirley Ardener and Jackie Waldren, for The International Gender Studies Centre, University of Oxford

ISSN: 1068-8536

Gender, Water and Development

Edited by
Anne Coles and Tina Wallace

Oxford • *New York*

English edition
First published in 2005 by
Berg
Editorial offices:
First Floor, Angel Court, 81 St Clements Street, Oxford OX4 1AW, UK
175 Fifth Avenue, New York, NY 10010, USA

Berg is the imprint of Oxford International Publishers Ltd.

Library of Congress Cataloging-in-Publication Data

Gender, water and development / edited by Anne Coles and Tina Wallace.—
1st English ed.
    p. cm.
    Includes bibliographical references and index.
    ISBN 1-84520-125-6 (pbk.) — ISBN 1-84520-124-8 (cloth)
    1. Water-supply—Government policy—Cross-cultural studies. 2. Water-supply—
Social aspects—Cross-cultural studies. 3. Women in development—Cross-cultural
studies. I. Coles, Anne. II. Wallace, Tina.

    HD1691.G43 2005
    333.91'22'082091724—dc22

                                                                        2005009254

British Library Cataloguing-in-Publication Data

A catalogue record for this book is available from the British Library.

ISBN-13 978 1 84520 124 1 (Cloth)
        978 1 84520 125 8 (Paper)

ISBN-10 1 84520 124 8 (Cloth)
        1 84520 125 6 (Paper)

Typeset by JS Typesetting Ltd, Porthcawl, Mid Glamorgan
Printed in the United Kingdom by Biddles Ltd, King's Lynn.

www.bergpublishers.com

Contents

Contributors

Gina Buijs is Assistant Vice-Rector and Professor of Anthropology and Development Studies at the University of Zululand, South Africa. She is the editor of *Migrant Women: Crossing Boundaries and Changing Identities* (Berg 1993) and has published on gender, ethnicity and development in Africa.

Felicity Chancellor has been closely involved in researching the socio-economic parameters affecting irrigated agriculture and the management of water in sub-Saharan Africa since the 1990s. Her main focus has been on smallholder irrigators – men, women and children – upon whom so many rely for food security. Her recent work highlights the need for equitable participation in the design and management of smallholder irrigation. She currently tutors for the Imperial College London Distance Learning Programme and chairs the British National Committee for the International Commission on Irrigation and Drainage.

Anne Coles is a research associate in the International Gender Studies Centre at Queen Elizabeth House (QEH), Oxford University. She was previously a senior social development adviser in the UK Department for International Development, leading on gender. Her career has combined university teaching (in geography and later development studies) with applied research and development practice.

Ben Fawcett is a water and sanitation engineer who has worked extensively with NGOs in many countries in Africa and Asia. Since 1996 he has led a masters programme in 'Engineering for Development' at the Institute of Irrigation and Development Studies, University of Southampton, UK, and manages research on social and institutional issues in the environmental health sector.

Sarah House is a chartered civil engineer currently working as a freelance water/public health engineer, based in the UK. She has had a professional interest in gender and equity in technical projects in development and emergencies for more than a decade, with a special focus on the practical approaches which can be used to consider gender and equity in institutions and in programme work. She has worked as an engineer and programme manager in Zambia and Tanzania and in research and consultancy in Ethiopia, Zaire (now DRC), Cambodia, Tajikistan, Azerbaijan, St Lucia and Ghana.

Anne Hutchings, ethnobotanist, author and compiler of the inventory *Zulu Medicinal Plants* (University of Natal Press 1996), has worked with healers and communities in health-related fields in rural KwaZulu-Natal. Since 2000 she has provided herbal treatment and monitored patients in an HIV/AIDS support clinic at a local state hospital. She is presently a research fellow in the Department of Botany at the University of Zululand.

Deepa Joshi, based at the University of Southampton, has her PhD and work experience in policy and institutional review of gender issues in rural water management and urban human settlements and sanitation in South Asia. She is especially interested in exploring the links between gender, poverty and livelihoods from an applied anthropology perspective, and is currently working on an urban water and sanitation research programme with Southampton University.

Michelle Moffatt lived in Nepal for eight years and worked with NEWAH as a gender consultant supporting the development and mainstreaming of a Gender and Poverty approach in the institution and programme. She also assisted the Government of Nepal and the Asian Development Bank in drafting a rural water supply and sanitation sector strategy from a gender and poverty perspective. She has more than ten years' experience working in project management in the UK and Asia. She has a Masters in Development Policy and Economics from Manchester University, UK.

Ben Page teaches geography at University College London. He worked, briefly, as a water engineer, before carrying out social research on the history of water supplies in Cameroon and Lagos from 1996–2003. He is currently working on an ESRC-funded research project studying development and the diaspora in Cameroon and Tanzania.

Umesh Pandey has been the director of NEWAH in Nepal since it was created in the early 1990s with the support of WaterAid, UK. He has been a strong supporter and advocate of the development and mainstreaming of a Gender and Poverty approach in his own institution, as well as externally. NEWAH is the largest national-level NGO in the water supply and sanitation sector, with five regional offices implementing projects in all regions of Nepal.

Shibesh Chandra Regmi is a development/policy analyst, with over twenty years of experience and expertise in issues related to development and poverty in developing countries, primarily in Asia, and in participatory approaches. His PhD concerns gender and development issues in water programmes in Nepal. He has been extensively engaged in the voluntary sector and since 2001 has been country director of ActionAid Nepal, an affiliate of ActionAid International.

Veronica Strang is a Professor of Anthropology at the University of Auckland. An environmental anthropologist, she has written extensively on water, land and resource issues in Australia and the UK, and is the author of *Uncommon Ground: cultural landscapes and environmental values* (Berg 1997), and *The Meaning of Water* (Berg 2004). She was formerly the Royal Anthropological Institute 'Urgent Anthropology' Fellow at Goldsmiths University in London, and a Senior Research Associate at the Pitt Rivers Museum in Oxford. Having been awarded an Australian Research Council 'Discovery' grant, she is currently working on a collaborative project investigating water issues in Australia.

Tina Wallace has combined academic work at Universities in Nigeria, Uganda and the UK (Aston, Birmingham and Oxford Brookes) with work within the NGO sector, especially with World University Service and Oxfam. She is currently a research associate at the International Gender Studies Centre at QEH, Oxford University and Open University, and works closely with NGOs in the UK and Africa on gender, strategic thinking and learning. She has published widely, most recently *New Roles and Relevance: The Challenge of Change in Development NGOs* (2002), a book edited with David Lewis and published by Kumarian.

Pauline Wilson is currently a freelance social development consultant with a special interest in gender and the use of participatory methods to promote more democratic development practices. She lived and worked in East and West Africa in the 1980s when she worked with Catholic Relief Services. During the 1990s she worked with ActionAid in a number of senior management positions with responsibility for countries in Africa and Asia.

Acknowledgements

This book is primarily the result of work undertaken by International Gender Studies (IGS) at Queen Elizabeth House, Oxford University, and the Institute of Irrigation and Development Studies (IIDS), Southampton University. Anne Coles and Tina Wallace, who were working with these institutions on water and gender issues, met at a dissemination seminar of the Southampton research in October 2002, and from there the decision to compile a book grew.

The Southampton University project was a research programme, supported by the UK Department for International Development (DFID, KAR project R6575). The study originally arose from discussions between the then Coordinator of Engineering for Development at IIDS, Alison Barrett, with colleagues at Oxfam GB and WaterAid, two UK-based NGOs working extensively in the development of water supplies in developing countries, and others concerned with the water sector. Their concern was the lack of gender analysis in water provision. The resulting studies and analysis are presented in three chapters of this book and were written by the two principal researchers for this project, Deepa Joshi and Shibesh Regmi, both natives of the countries where they studied from 1998 to 2000. Their work was supervised by a team comprising: Ben Fawcett, who became Coordinator of Engineering for Development at IIDS after the study was conceived; Nicky May, an independent consultant in gender and development; Simon Trace, Head of International Operations with WaterAid; and Tina Wallace, co-editor of this book. Through this research contact was built up with WaterAid and with water agencies in Nepal and elsewhere; this directly led to the commissioning of other chapters for the book. Sarah House, Michelle Moffat and Umesh Pandey, and Tina Wallace and Pauline Wilson, who undertook review work with WaterAid, which informed their chapter, contributed these chapters.

The genesis of this book within IGS was a seminar series convened for the Centre by Janette Davies, Shirley Ardener and Anne Coles in the Michaelmas Term (from October to December) 2000. The papers presented there and revised for inclusion in the book were written by Felicity Chancellor, Anne Coles, Ben Page and Veronica Strang. The chapter by Gina Buijs, a former member of the Centre, and Anne Hutchings was later commissioned by Anne Coles who knew of their work in South Africa. Other speakers in the IGS seminar series presented material that has informed the thinking and analysis behind the book, especially the work of Jeffrey Davies and Janette Davies, who presented a paper addressing the need for better understanding and monitoring of water sources and replenishment, something missing from current government and NGO work on water supplies. Their paper 'Water Beneath Their Feet: Aquifers that Supply the Needs of Rural Africa' also explored the various ways in which water can be drawn from the earth in times both of plenty and drought. Jacqueline Waldren chaired a panel on 'Development and tourism: consequences for water resources', an important theme she has developed since in a paper, soon to be published elsewhere. In addition, two Visiting Fellows participated. Grace Ave, from Ghana, talked on ownership and sustainability of rural water schemes in Ghana, and the implications for female participation, while Nuria Roldan from Madrid spoke on the management of water in the island of Gran Canaria. Judy Mabro presented 'Naked Emotion: the Ambiguous Pleasure of the Public Bath', part of a book in preparation based on her research in the Middle East. Lastly, Sally Kendall, spoke of the early stages of life and babies' first introduction to liquids, including breast milk. It is our expectation that these interesting materials will soon be forthcoming in other publications.

Thanks are due to all those whose work in Southampton and Oxford Universities and beyond contributed to the research, analysis, and understanding that have informed the writing of this book.

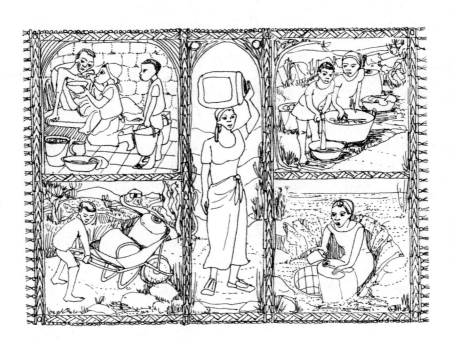

Drawing by Anne Hutchings, illustrating water-related problems experienced by women and children in rural KwaZulu-natal (see Chapter 10, this volume). They are shown nursing the sick, undertaking laundry in the open, and transporting and collecting water from hand-dug wells. The background motif is from local grass place mat weaving, a crafty also restricted by drought.

1

Water, Gender and Development: An Introduction

Tina Wallace and *Anne Coles*

Invisible and Invincible Boundaries: The Context

This book is about one of the world's most precious resources, water, and the challenges of enabling all people to enjoy the benefits of reliable water supplies. It explores, through a variety of lenses, the role of gender relations, in different cultures, in water production and delivery and in the use of water within households. Deepening understanding of the ways in which gender shapes who has control of water, who gets access, the different needs and positions of women and men, and the issue of rights, is crucial for development. The book attempts to straddle the invisible and yet often invincible boundaries between practitioners and academics that all too often inhibit broader discourses; it brings together theoretical analysis and empirical evidence. While practitioners are concerned with immediate problems and workable solutions, academic researchers, taking a longer-term perspective, explore process and causality (how things work and why) to extend knowledge. This book warmly embraces both. The authors include researchers, development workers and those straddling the academic–practitioner divide. In this way the book implicitly and explicitly addresses many of the issues raised by current work in the water sector, with the intention of contributing to more informed ways of tackling them.

The contributions are diverse, but all are based on original thinking and research, bringing new gendered information and fresh analysis to the debates around water. The approach is interdisciplinary, drawing on the traditions of history, sociology, anthropology, geography and engineering. Adopting a historical perspective, the book spans many countries and cultural settings, concentrating on rural areas where services are most lacking. While the specificity of each situation is crucial to understanding gender relationships and their implications for access to and use of water, the constant throughout is that water is gendered in every society.

Each chapter stands alone as a major contribution, but a range of issues emerges from the collection that is critical for those currently grappling with water and gender issues. While the international development discourse of current water policy and practice formally acknowledges the importance of gender in water supplies, the results are, with exceptions, disappointing. There is a need to delve deeper into the gendered nature of water, and into the historical reality that gender has shaped water management over centuries, in order to understand what is required to successfully turn existing gender commitments into good practice and ensure that water systems meet the aspiration of 'water for all'. This book, through the different chapters, offers the knowledge and experience of its contributors to extend thinking and analysis; to throw light on recent development trends; and to offer alternative approaches.

This introduction serves as a guide to the rest of the book and begins with an overview of the development discourse on water, a review of current thinking on gender and development, and a synthesis of how these two strands interact in water policy and practice. Each chapter is then introduced to enable readers to see where and how different issues are addressed within the book, and some critical cross-cutting themes drawn from the chapters are highlighted to guide the reader through the many contributions.

Water Worldwide

Global Water Statistics

Water interventions in development have changed over time since the 1950s, when the central role of water in public health was identified during the 'health for all' campaign. They have always been a central part of the development agenda and water was accepted early on as a 'basic need'; in 2002 the UN defined it as a 'basic right'. Yet, in spite of decades of commitment to meeting local water needs, global water coverage varies significantly between regions and continents, and 1.1 billion people remain without access to acceptable water supplies.

Only 50 per cent of the world's population has access to piped water; the figures are only 4 per cent for Africa, 12 per cent for South Asia and 8 per cent for South-East Asia, regions that are the primary focus of this book. The Task Force for the water Millennium Development Goal 'to provide access to safe water and basic sanitation' states that 'for every person in urban areas there are six people in rural areas without improved drinking water sources' (WHO/Unicef 2004: 22). The situation for sanitation is worse, with 2.6 billion of the world's population without any sanitation services even though it is widely accepted that 'water and sanitation are among the most important determinants of public health' (Director, WHO, August 2004).

The Development of Water Policies

The early development focus was on supporting centralized, government-run public sector departments, and water was no exception. Male engineers, largely following blueprints drawn from water provision in the wealthier countries of the North, ran projects. While women were understood to bear the burden of poor water supplies, they did not feature in water policy or delivery systems in the 1950s and 1960s. The focus shifted in the 1970s and 1980s, when the obvious fact that women were the managers of domestic water, and usually the carriers of it as well, was recognized. As clearly identified water users they were seen to have the knowledge required to maximize the value of water supply improvements and they were formally accepted as a constituency in water development and management. The water sector was therefore among the first to recognize women's potential contribution to development. Key agencies committed themselves to meeting the needs of women and promoted their participation in water supplies and water management (INSTRAW 1985). These two decades also saw a widening of concerns beyond the provision of drinking and domestic water, to include environmental issues around water, and women were often defined as the guardians of that environment.

However, the United Nations, during its development decade on water, 1981–90, again focused only on drinking water and sanitation and 'aimed for clean drinking water and sanitation for all by 1990' (Joshi 2002: 49). The involvement of women was promoted, largely for economic reasons: women's participation was expected to increase the efficiency of water projects, because of their interests in achieving reliable supplies of domestic water. They were trained as caretakers, health educators, motivators and hand pump mechanics (Ghosh 1989). They were also expected to be active in income generation, to use the time saved from collecting water productively (INSTRAW 1985). The costs, constraints and barriers to their involvement were not seriously discussed.

The 1990s witnessed ambiguities in water policy, generated by a number of different development trends (Joshi 2002: 59). These included the rejection of the role of the state in public provision, which was seen as having been costly and inefficient throughout the 1970s and 1980s, and the cuts in public sector spending imposed through structural adjustment programmes. The market, often defined as both private sector and non-governmental actors, was to provide basic services instead, and government's role was to enable and regulate private sector provision, and not to provide water directly. Water provision could best be met within the neo-liberal economic agenda. Privatization, decentralization and demand management (enabling consumers rather than suppliers to set the levels of water supply, according to their needs and ability to pay) became the central modalities (World Bank 1993). There was a global commitment to 'water for all', and community participation, gender (usually meaning the inclusion of women) and empowerment were all recognized as essential to achieving this.

These approaches were formally endorsed in 1999 when over a hundred countries agreed to the Dublin principles. These affirmed water as 'an economic good', which is finite and vulnerable but essential to life. They prioritized privatization, water pricing and cost recovery to ensure project efficiency, while emphasizing participation at every level. Women were defined as central in providing, managing and safeguarding water, a role reinforced in Agenda 21, drawn up at the Environmental Summit in Rio the same year. The potential disjuncture between the economic and social goals, and the complex challenges of water provision in contexts of inequality, were not widely discussed within the sector (Joshi 2002).

Most recently, the focus has been on the role of water in poverty alleviation and women's empowerment. Water is defined as a right; the current World Water Vision is that almost everyone will enjoy safe and adequate water and sanitation provision by 2025 (World Water Council 2000). This is to be achieved within the economic approach aimed at achieving the full-cost pricing of water services to ensure continued provision.

The current policies do not seem to reflect evidence from the field. There is little good research into the social issues involved in the provision of water in poor countries. Most formal evaluations are superficial, counting heads and tap stands; few set out to analyse the impact of water projects on social relations and, consequently, who does or does not have access to water, how it is used and who benefits. There has been little systematic learning about how projects work and about what happens to improved water supplies after five to ten years. Although the specific circumstances are important, there is still too little exchange of information on how to make the most of women's expertise when developing new water points. It remains relatively unclear what approaches help to promote positive change for women in different contexts; history and experience show, however, that achieving improvements in the status of women is a long-term process that will continue beyond the working lives of the present generation.

The production, distribution and consumption of domestic water for drinking and household use tend to be largely separate from other aspects of national fresh water provision. Yet domestic water accounts for only 10 per cent of consumption and needs to be understood within the larger context. The bulk of water is used for agriculture (65–70 per cent) and industry (20–25 per cent); much of this may be wasted through overuse because pricing structures often favour agricultural and industrial users above smaller household consumers and because the private sector 'demands increasing consumption while contributing little to the conservation of resources' (Barlow and Clarke 2002: 126). This issue of conservation is critical, yet little attention is currently paid to it throughout the water sector. As provision shifts from governments and international geological bodies to private agencies (both for-profit and NGO), a vacuum is appearing over whose responsibility it is to map and monitor the geology and rates of replenishment of water globally (Davies and Davies 2002).

The Evolution of Gender Theory: Linking with Development Policy and Practice

In the 1970s, understanding of the roles and needs of women and men, their interrelationships and the power differential between them began to deepen. The critical need to address these in development was recognized. Discussions were influenced by feminist theory in Europe and the USA. Gender theory in development evolved out of thinking by key writers such as Oakley and Boserup in the 1970s, who distinguished between biologically determined sexual attributes and gender, separating out the ascribed roles and expected behaviours, shaped by culture and history, given to men and women from birth.

The early 'women in development approach' (WID) shifted the focus of development from women as passive recipients of projects towards women as actors in development. Women's roles were recognized beyond the domestic sphere, in production, social and community work (Moser 1993): previously women had been largely recipients of welfare approaches to development. Over time analysis deepened to include women within their networks and relationships, their positions within households, communities and at national level, and how their status was maintained and reproduced.

There was a growing concern to change existing patterns, which showed high levels of female subordination. Women's empowerment was introduced: how to enable women to break their shackles (Mukhopadhya 1984) and fulfil their potential. For some this was an economic imperative, for others it was a question of women's human rights. To achieve change, recognizing the specificity of women's condition and position in each culture and society was essential. Gender roles and positions could not be understood without careful analysis. Knowledge developed around the ways in which class, race, wealth and religion also shaped women's experiences and society's expectations of them (Kabeer 1994, Molyneux 1985). Women scholars and activists from the South challenged 'white dominated' understandings of feminism, and contributed to the recognition that gender relations were deeply rooted both culturally and historically, and that addressing them required knowledge of each context (Kandiyotti 1988). Women were no longer defined as a homogeneous interest group, although gender relations and the dominance of patriarchy were universal (Whitehead 1979).

The work of gender academics and gender advisors and activists cross-fertilized, although not always very easily (Wallace and March 1991). New issues, previously overlooked in women in development agendas, emerged from field experience of working with women and gender issues. There was a shift away from working with women alone to address their problems, especially their subordination and lack of power in many societies. The new gender approaches were designed to address inequalities between women and men, and worked with men as well as women, in the community and within institutions beyond the community level because power inequalities were evident at the meso and macro levels, as well as at household level (Elson 1991). Work took on board both the causes and the effects of women's relative powerlessness and approaches were promoted to enable them to have access to previously unacceptable forms of interaction and communication. These approaches included the recognition of domestic violence against women as a development issue, accepted in a UN convention (CEDAW) in 1979, women's need for control of their own bodies, and the importance of helping women to find 'their voice' and to represent themselves in public fora. Women's right to equality, central to gender analysis, became enshrined through the International Covenant on Economic, Social and Cultural Rights, which came into force in 1976.

The dominant narrative of development, which was predominantly male, was challenged in academic circles and in practice. International and national policies, and organizations at every level down to the household, as well as cultural and re-ligious norms in different societies, were seen as perpetuating gender inequalities. Work to tackle these inequalities was taken up internationally in 1975–85 with an international UN decade for women and a conference in Nairobi to celebrate the end of that decade. Policies and principles promoted there were consolidated ten years later in Beijing, in the comprehensive Platform for Action, which most governments endorsed. This has been monitored through 'Beijing plus five' and will be reviewed in 2005, under 'Beijing plus ten'. So far the results are disappoint-ing. Interest, especially from donors and Northern governments, has waned since the mid 1990s (Painter 2004). Some significant gains have been made, however, in other forums, especially in international health at the Cairo Conference on Population and Development in 1994 where women's rights to control their sexual and reproductive health were agreed, following massive struggles and opposition (UN 1994).

Gender theory arose out of, and informed, different international social move-ments for change for women. It was intended to transform both male and female roles, leading to change in institutions and individuals. Only in this way would it ensure that women, and indeed others subordinated by class, caste, poverty and religion (previously ignored and marginalized) could have a voice. It demanded that women be treated as equal participants in the development process, and that ways be found to enable this to happen.

Gender Frameworks and Gender Mainstreaming

Gender theories were complex because gender relations are shaped by history and geography, and are rooted in different social, cultural, economic and political structures. This complexity led to a felt need within the development sector for more manageable ways of explaining and addressing gender issues that could be easily understood and implemented alongside existing ways of working. Frameworks for understanding women's roles and interests, their responsibilities and social position, were developed for use by development practitioners (Moser 1993, Molyneux 1985, Kabeer 1994, Levy 1998). Methods for analysing the hierarchies of women's participation were developed (Longwe 1991) and practitioners introduced gender analysis into Participatory Rural Appraisal techniques (Welbourn 1991, Guijt and Shah 1998). Institutional 'packages' were developed 'to mainstream gender', which included the adoption of frameworks, training, disaggregated data collection, the recruitment of gender specialist advisers and sometimes the establishment of gender units.

Over time gender became increasingly regarded as a technical 'problem' to be solved by applying the tools provided by these frameworks, with the risk of oversimplification and eventually distortion. The experience, analysis and commitment associated with the rise of gender theory and practice were glossed over, the challenge was blunted, and the concept of empowerment often lost its focus on social transformation. Many voiced their concerns about this 'mainstreaming' approach:

> ... gender analysts fear [that] the dangers of mainstreaming gender as practiced currently misinterprets and reduces gender to women/men – stripping away the consideration of the relational aspects of gender, power and ideology and of how patterns of subordination are reproduced ... while valuable insights and empirical evidence are provided, gender issues are delinked from the feminist transformatory purpose. (Joshi 2002: 42)

> To retain its dominance, the dominant culture must not only adopt a mask of progress, but also incorporate some minor elements of change. It must absorb, assimilate ... new demands, so that it can remain dominant. I deliberately use the language of race relations, since parallels have been drawn between the conditions of race and gender, and although they differ in important ways, there is a certain commonality in the way the dominant culture deals with difference and perceived threats ... so, if our suspicion is correct, bringing gender into the mainstream will result in absorption, dilution, being carried away. (Walker 1993: 2)

Gender became the responsibility of everyone, but sometimes, in practice, of no one. Gender mainstreaming work focused primarily on communities, less on gender hierarchies within organizations, and provided little space for challenging dominant development values. Most institutions attempting to deliver gender mainstreaming remained hierarchical and relatively insensitive to gender internally.

The use of gender frameworks certainly enabled modest improvements for some women, and gender specialists continue to work on improving them.

However, the preliminary gender analysis essential for gaining an understanding of the local cultural context was often skimped on, and the radical rethinking of purpose and process to ensure that women, as well as men, can influence the concept and design of development initiatives was often missing. As a result, while gender is a term that is widely used, it is often poorly theorised and ill defined, and gender policies are weakly implemented.

How is Gender Taken into Account in Water Policies and Practice? A Critique

References to women come and go. Women are defined as essential providers and users of water. They are expected to play multiple roles in the provision of water, sanitation, improved health, and increased productivity – because they are effective instruments in water provision, 'the cost effectiveness and positive impacts of a gender approach in the water sector have been amply demonstrated' (World Water Forum 2003: 1). Most policies tie women tightly to the socially ascribed gender roles they play in relation to water in most poverty contexts.

Within the present economic paradigm, women's social and cultural roles and status are poorly analysed and their ability to pay for water is often simply assumed. Their subordination and the consequent barriers to their active involvement in influencing water programmes are barely addressed. There is limited attention to women's rights to water and what these would mean in practice in poor communities where women's status is often very low, although the need to do more to realize women's empowerment is acknowledged (World Water Forum 2003).

Gender is often equated with 'women' in the water sector. Where attempts are made to go beyond this, gender is increasingly conceptualised simply as 'women and men'. Gender work is expected to ensure that both women and men are involved and benefit and often the focus is firmly back on men (Joshi 2002, Gender and Water Alliance 2000). Gender concepts of power inequality and the importance of transformation to ensure that women achieve positive and lasting changes in status are largely absent.

The international water sector, including the World Bank, DFID, Global Water Partnership and World Water Forum, endorses privatization as the key approach to providing water for all. While the debates continue, often heatedly within some NGOs and think tanks that promote alternatives based in participation and community ownership (TNI 2004), the focus on privatization, with its emphasis on the economic value of water and the need for full cost recovery, has affected work right across the water sector. Yet the concept of private ownership and control fits ill with deeply held values that water is a gift of God, a public good, and a human right. From a gender perspective, evidence increasingly shows that the private sector is not regulated or bound by the global commitments to gender equality, women's empowerment and providing water for all. Water is a highly lucrative sector and water companies have limited interest in meeting the needs of the poor, including women (Polaris Institute 2004, Aegisson 2002). Access to water is based primarily on an ability to pay, not on need or rights. While governments are charged with regulation of the private sector, they lack capacity and political influence over transnational companies (Gutierrez et al. 2003). The poor have no power over these companies and their participation is not essential to the work.

Many predicted the inability of 'for profit' water companies to meet the needs of the poor in very poor countries; the evidence now supports their analysis: 'privatisation [c]ould not deliver the promised efficiency and improvements to access to clean drinking water for the poorest. Escalating prices and non-fulfilment of promized investments were common features of privatisation' (TNI 2004: 1). 'Privatization is one aspect of the world water crisis that is having a deeply negative impact on the livelihoods of women' (Grossman et al. 2004: 1) The World Bank still promotes privatisation and cost recovery 'despite evidence that such policies reduce access, raise the price of water for the poor, exacerbate inequalities and reduce local control' (World Bank Watch June 2004); the European Union's new water facility (2004) also supports private sector water investments through subsidies from public funding and development aid.

The water sector, like all development sectors, currently delivers support to countries through direct budget support to national governments. How the money is to be used is agreed through the Poverty Reduction Strategy Plans (PRSPs), which aim to ensure aid reaches the poorest. These plans, while officially owned by developing country governments, follow the parameters set out in structural adjustment programmes and may be heavily donor-led. Research has shown that these plans are light both on water provision (DFID 2004a) and on gender analysis: they largely lack commitment to women's needs and rights (Whitehead 2003).

Since 2000, the UN Millennium Development Goals (MDGs) have also been key in shaping development thinking and practice. These set clear international development targets and have generated a great deal of political support worldwide; they serve as a focus for assessing global progress in meeting a range of basic needs. One of the targets relates to domestic water provision; three relate directly to women and gender. However, the water goal does not incorporate a gender perspective – its focus is on coverage rather than access or equity. The links between the different development targets are weak, and issues such as how girls' roles in fetching water may affect their access to education – a key to fulfilling their potential – are not addressed (DFID 2004b). The data collection and monitoring for the 2005 midterm review lacked analysis of who has access to, and who benefits from, improved water provision, whether at country, community or household level. The significance of this gap for assessing progress on reaching the poor is now recognized: 'from now until 2015 greater efforts must be made to reach the poor and those in rural areas, whose deprivation is hidden behind national averages' (WHO/Unicef 2004). There are plans to disaggregate data on water by gender, poverty and region in future, something gender analysts have long highlighted as essential, although the complexities of gathering such data in comparable ways across countries are a cause for real concern.

The Issues Addressed in the Book and Their Relation to Current Practice

Many of the chapters take up and examine in detail the issues raised by a gender analysis of water and current policies and approaches. The chapters are cross-cultural, with case studies from sub-Saharan Africa and southern Asia, as well as from the UK. There are two pairs of interrelated chapters. For Asia, the evaluation of three water programmes in Nepal (Regmi) is followed by a more optimistic report on how one of the organizations concerned, the NGO, subsequently adopted a gender and poverty approach to its work (Pandey and Moffat). There is a case study of a World Bank project in India, which illustrates Joshi and Fawcett's theoretical discourse on the interrelationship between Hindu philosophy, water and gender. The chapters from Africa have a broader coverage both in terms of countries and themes. Page deals with women's historically proactive role in the development of water supplies in Cameroon, while Hutchings and Buijs provide a contemporary study of women adjusting to the AIDS pandemic in a water-short area of South Africa. Coping with marginal water supplies is the theme of a chapter that examines the complex interrelationship of ethnicity, gender, water and ways of life in the Sudan (Coles). Chancellor stresses the importance of sensitive and participatory gender analysis in smallholder irrigation schemes in southern Africa, while House provides a case study of how a gendered approach was incorporated into a water programme in Tanzania.

In contrast, Strang's study comes from Britain. A sweeping historical survey, showing how women's relations with water changed as social systems and technology became less locally based, it has implications for water providers globally as provision becomes more centralized. Wallace and Wilson consider the development of gender policy and practice in UK NGOs, focusing on one of the most significant international water NGOs.

While none of the chapters is based on a case study of privatization per se, the elements associated with this approach, especially cost recovery, reach deep into several of the studies. The chapters show that most approaches lack an understanding of the depth and breadth of gender inequalities, and how these are rooted in history and geography, in the contexts where water projects are delivered. Concepts such as community and gender are shown to be ill defined; it is demonstrated that a homogeneity of community needs and interests is often assumed and that gender differences and inequalities – along with other dimensions of social marginalization – are often not assessed. The studies illustrate how gender roles are often largely accepted as given, rather than challenged. We see that while policies highlight women's empowerment, implying strategic and transformational change, often the practice is focused on meeting women's practical needs as domestic water users. The commitment is to provide water for all, yet it is shown here that how the needs of women and the poorest are to be met remains unclear. The rhetoric of poverty, gender, equality and empowerment is strong, but the policy instruments and evaluation procedures belie these commitments.

The book begins with Strang's historical and ethnographic story of a river in southern England. It traces a broad narrative, observing the social, conceptual and technological transformations that have, over time, enabled patriarchal and elite institutions to appropriate previously collective water resources. As new religious and scientific concepts reframed the meanings encoded in water, there were important – and related – shifts in gender and power relations, which impacted negatively on women. Later, enclosure of water resources and developments in material culture largely divorced local communities from water provision, as regional and central authorities assumed the task. Strang highlights the analytic importance of considering the long-term patterns of change that have created particular socio-economic, political and material landscapes – cultural landscapes in which the control of water is highly gendered and increasingly contested. As water supply and distribution become increasingly controversial internationally, it is plain that the long-term patterns revealed by the historical ethnography described here are echoed in many countries around the world, and that an understanding of culture, myth and history is vital to analyses of current conflicts.

Like Strang, Joshi and Fawcett explore the role of religion and myth on water use – describing their continuing relevance for water access and use in India at the present time. They discuss the historical development of Vedic philosophy, which underlies Hinduism as practiced today, emphasizing the spiritual role played by water as a means of purification but also as a substance which is readily polluted. The caste structure perceives lower-caste Dalits as permanently impure, while bodily secretions render women temporarily impure. Importantly these beliefs determine access to water; this is illustrated in a series of stories from hill villages in Uttaranchal in northern India. The interrelationship of caste, class, power, women's inherent inferiority and ideals of purity results in women, who are the fetchers of water, having unequal and disrupted access to water sources. Dalit women are doubly disadvantaged – both as women and as members of the lower caste. Thus, despite government efforts to abolish the caste system, the pervasiveness of traditional Hindu belief in unequal social order results in discriminatory access to water.

Much of the contemporary literature on women, water and development emphasizes the role of women as consumers. In contrast Page deals with women's historically proactive role in the development of water supplies in Cameroon. His chapter, broadening the concept of 'water production' to include both the material process through which water is accessed, transported, treated and distributed, and the discursive process through which water becomes meaningful, shows women's important participation in this production. In their basic role as fetchers and managers of domestic water, women were also involved in water technology, water institutions and water politics. Although formally powerless, women used authoritative traditional means of protest to lobby men and institutions successfully to achieve concrete improvements in local water supplies. The chapter argues that failing to take into account fully women's historic involvement in water production may actually result in modern development programmes which effectively disempower them.

Coles' contribution is based on a particular historical moment: both the research and initial (though not subsequent gender) analysis were undertaken in the early 1960s in a semi-arid area of Sudan. The area had varied sources of water and a diverse immigrant population. The chapter emphasizes the limits that the physical environment imposed and still imposes on water availability in many parts of the world. Political considerations, technology, and economic constraints affected how much of the potentially available water could be exploited. But the utilization of that water was determined largely by ethnic and gender considerations. Water was a social construct; perceptions of water adequacy and water requirements were cultural; they were related to ways of life and to economic aspirations. Gender roles and responsibilities relating to water, shaped by religious and tribal values, influenced these perceptions, as well as economic activities and settlement patterns. The response to an unexpected drought, which is vividly portrayed, emphasized the fragility and marginality of water supplies in relation to population growth, service provision, living standards and overall development.

In his chapter, Regmi evaluates the extent to which three very different organizations providing domestic water in Nepal addressed gender issues. Using the standard tools of gender analysis, his approach is thorough, systematic, and comprehensive – and his findings are worrying. Some of the problems which he identifies seem to be all-too-typical of development agencies elsewhere, as other chapters show. But in Nepal the situation was aggravated by general acceptance of deeply held feudal and patriarchal values and norms, which permeated the organizations concerned and resulted in failure to challenge gender stereotypes, making it hard for women staff to be treated as equals and for gender initiatives to be more than sporadic. Additional practical barriers, notably lack of qualified women and social taboos on women travelling, impeded equitable recruitment and staffing policies. Inefficiencies resulting from failure to involve local women in decision-making were clearly visible in the field, though seldom discerned by staff in the organizations concerned. The author advocates more holistic and transformational gender policies to change attitudes and behaviour – albeit within a realistic understanding of local society.

Enlarging on the organizational issues involved in adopting gender approaches, Wallace and Wilson explore those confronting international development agencies involved in improving domestic water supplies for the poor, and how they manage a range of conflicting expectations and demands from donors, staff, partners and local communities. The chapter is based on a case study of WaterAid, a large international NGO, initially funded by British water companies specifically to provide water to ill-served communities in the South, but the findings apply broadly. Accustomed to working on technical aspects of water supplies in a situation where there is universal affordable provision, the male-dominated industry initially failed to appreciate the relevance of the social issues involved in accessing water in much of the world. WaterAid now promotes community participation, having learnt from experience the importance of understanding the social context essential to ensure the effectiveness of its engineering efforts. But the desire to scale-up operations (in terms of water points provided) and reach ambitious coverage targets in short time frames results in on-going tension between the 'soft' and 'hard' aspects of development. There is ambivalence over the extent to which time should be spent on equity issues – particularly gender. With cost recovery accepted as a basic principle, further attention is needed to issues such as how poor women pay. There is no overall gender policy. Nevertheless, some country programmes do give gender serious attention, and gender equity has its champions, mainly women in middle management. Recent changes have resulted in greater internationalization of senior staff and board members but women in top positions remain a minority and decisions about working to address gender inequalities within the organization and in development are left largely to individuals.

Returning to the Himalayan foothills described earlier, Joshi rehearses gender theory to critique the World Bank's and government of India's SWAJAL initiative. This project was lauded for its success in providing improved water supplies and in achieving community participation. However, Joshi's analysis suggests that by incorrectly assuming 'unitary, holistic and altruistic' communities the result was to reinforce existing social inequalities. Women and Dalits might be formally included in committees as intended, but powerful social hierarchies ensured that they were effectively still voiceless. In some cases the very poor, officially targeted by the programme, became relatively or even absolutely worse off in terms of water provision. Caste and gender discrimination are so deeply embedded in people's values – and are so instinctive – that even local NGOs fail to be sufficiently inclusive in their approach.

It is good for this book to have the added dimension of water for irrigation, especially as women's vital role in African agriculture has been recognized at least since the 1970s. Of particular interest is Chancellor's detailed analysis of the interrelationship between the technical, the social and the economic aspects of smallholder irrigation projects (in different contexts see also Coles and Strang). The chapter is not only about women, although Chancellor is clearly their champion in addressing male bias, but is also about the respective and interrelated roles and responsibilities of both men and women. The author presses for an understanding, not only of culture but of the dynamics of demographic and social changes.

Chancellor believes that a sensitive, participatory approach is essential to ensure that the design of smallholder irrigation meets the needs of the prospective irrigators. Achieving such participation is not easy, notably when views are sought about processes that may be unfamiliar and when women are unable to speak out in public. But this is a practical chapter, based on an applied/operational research project in southern Africa, which provides guidance and advice. If you think that a hoe is just a hoe or a pump just a pump, read on – both are infused with gendered meanings!

Most papers dealing with water and health focus on issues of water quality and quantity or the role of water in disease transmission. Hutching's and Buijs' chapter is different: it demonstrates the importance of water in the care of the chronically sick. The sick are people with HIV/AIDS so that this case study from South Africa has great relevance and immediacy today. Globally, many of those with HIV/AIDS are in rural areas, where services are poor and water supplies problematic.

This is a story about women in a patriarchal society but one where men are largely absent through death or migration. The story is partly told in the words of the elderly women in the community, particularly those with responsibilities for caring for family members who are terminally ill. Their fortitude is remarkable; their ability to cope is an example of women's strength, in a context of vulnerability and traditional lack of autonomy. Their voices illustrate their preoccupation with water, and with the greatly increased need for water of patients suffering from AIDS in a situation where water was becoming increasingly scarce and its collection increasingly time-consuming. The area has experienced environmental degradation, which has reduced water availability and the capacity of the land. In remembering a time when greater harmony with nature prevailed, the women reveal a finely tuned local understanding of ecological patterns.

NEWAH was one of the NGOs whose gender approaches (or lack of them) were assessed by Regmi. From 1999 onwards, however, NEWAH, with strong leadership from the top, adopted a gender and poverty approach, subsequently funded by the British government's ministry DFID. This chapter is cautiously optimistic and written by Pandey and Moffat, who are respectively the NGO's director and former gender expert. Relying partly on NEWAH reports and an evaluation of the Gender and Poverty (GAP) approach, this is a finely documented account of the process of carrying out the strategy, the problems met and the progress made in the inhibiting cultural context of Nepal. It is the only chapter to discuss the sanitation components that typically accompany water projects.

The chapter emphasizes the sustained commitment required if equity and empowerment are to be achieved. Changes were and still are needed, both internally within the implementing organization and externally in how the programme is delivered, particularly in interaction with communities. Among the lessons learnt are the need for long-term, sensitive planning and for a flexible approach, which listens and responds to local needs and wishes in terms of the detailed design and siting of facilities. The authors show that it *is* possible to address equal access to improved water regardless of caste (even if caste discrimination cannot always be tackled in the short term), and that the approach *can* contribute to the empowerment of women by modifying gender roles, so that women play a more active part in public and men a greater role in domestic life.

House, who was a programme manager implementing a water programme in northern Tanzania, shows the day-to-day reality of trying to adopt a gender and equity approach in the delivery of improved water supplies in the field. Here are multiple voices: House's own professional reflections, documented at the time, those of the programme's NGO staff and finally the voices of people from the communities served. The author provides a balanced view of the preliminary results of the programme but it is these last voices, of women who have felt empowered and of those who are still sceptical, that provide evidence that it *does* seem possible to improve the status of at least some women through the process of developing water supplies, thus reinforcing Pandey and Moffat's conclusion.

House also identifies barriers to achieving and maintaining a gender and equity approach: lack of commitment of senior management (also discussed by Wallace and Wilson); the uneasy position of gender activists, certainly needed but uncertain in their effectiveness (it was some local female managers who were champions of gender equity); and the vulnerability of gender activities in the absence of clear policies. Importantly from a practitioner's perspective, she highlights a lack of really practical advice, of advice on 'what to do when women don't come to meetings'. Analysing the different perceptions of participation held by development professionals, she emphasizes the importance of understanding complex local hierarchies of power if projects are to include the vulnerable members of society in the development process.

Critical Issues for Future Work on Water and Gender

The following cross-cutting issues emerge from the book as a whole and seem, to the editors, critically important for addressing the issues raised in the context of developing a more coherent gendered approach to water.

Understanding the History of Gender Gives a Better Framework
The deep-rooted nature of gender inequalities and their relationship to water are well illustrated in several chapters, yet targets are developed which are usually short term and changes in water management and behaviour are expected to occur quickly. Lessons from both the historical and geographical studies could enhance long-term thinking about how best to address the issues embedded in the water–gender nexus.

Addressing the Reality of Gender Inequality at Community Level
Gender inequality is present in every context, and is cross-cut by class, caste, religion, geography, and wealth. Women and men are not homogeneous groups and cannot be treated as such; various social inequalities mutually reinforce each other in each context. Simple frameworks and approaches cannot capture the specificity of these social inequalities, yet addressing them is essential both for social justice and to prevent conflict and dysfunction at the community level over water. This is also a major challenge for modern-day providers – how to provide basic services in ways that address these inequalities in practice and are genuinely inclusive. Three chapters especially explore new approaches, with interesting learning for the future (House, Pandey and Moffat, Buis and Hutchings).

Gender within Organizations: The Importance of Leadership and Training
Gender is an issue affecting organizations as well as communities. Within organiza
tions, leadership is identified as critical, and strong leaders committed to gender do make a real difference to addressing gender inequalities internally and externally (House, Pandey and Moffat, Regmi). The cases show that much innovation comes in practice from committed women, who are often middle managers; however, while they bring new approaches to the gender work, they are often unable to shift the overall focus of the organization (House, Wallace and Wilson). Where senior management, usually predominantly male, does commit to gender work, this can change the whole organizational culture and approach (Pandey and Moffat).

Staff training is also key to working in deeper ways with communities and managing the reality of the diversity of groups *within* communities (Coles). Training cannot be one-off. It needs to be ongoing and embedded (Regmi, Pandey and Moffat). However, this often conflicts with the organizational cultures of many agencies, where the priorities are extending water coverage, meeting targets in short time frames and solving technical problems. Often, as a result, resources are provided in ways that do not enable the weakest to benefit long term (Wallace and Wilson).

Gender Relations Change Over Time and are Affected by New Technologies

Gender relations change over time, and respond to the introduction of new technologies and ways of working. While technology is often seen as gender neutral, experience shows that technology alters existing gender roles and responsibilities, and often favours men (Chancellor, Strang). Women and those marginalized by caste, class or poverty may lose the assets they previously controlled when new technologies are introduced (Joshi). Attention has to be paid to how technology can meet the needs of those it is intended to serve, ensuring it is introduced in ways that do not reinforce or further extend gender inequalities.

Women as Agents of Development: The Importance of Their Participation

While there is much discussion of rights-based approaches to the provision of basic needs, and the importance of women as agents (not recipients) of change, current practices still overlook what these mean in practice. Women are often involved not only in the consumption and use of water, but also in its production (Page), yet this is often ignored. In spite of the providers' claims to promote women's empowerment, work to achieve this is frequently omitted. Failure to recognize women's active roles often results in misunderstanding of their needs and interests, and their further disempowerment.

Participation is a critical tool for overcoming this, and aims to involve women and men in design through to evaluation; it can, however, be relatively meaningless if it is implemented as a 'quick fix' or tick box exercise. It has to be promoted carefully, ensuring that those who are usually excluded are involved through building their confidence and developing new skills, as well as working to ensure that men will allow women to participate (Chancellor, House, Joshi, Pandey and Moffat).

Gender Rhetoric and Reality: Gender is not an Optional Extra

The major aid providers dominate the development discourse and it is unclear what has become of the feedback loops that should enable policy and associated project 'packages' to be rejigged to reflect field experience. There seems to be an increasing separation of rhetoric from reality and current approaches appear poorly related to experiences in organizations and communities. How much flexibility is there for local culturally sensitive approaches to flourish and how much negotiation is possible within these dominant paradigms? Successful initiatives need to be shared, learnt from and replicated, but mechanisms for this appear weak. The current structures raise the question, to what extent are providers currently focusing narrowly on water provision, subverting, themselves, their aim of bringing sustainably improved water resources to ill-served communities?

Attending to issues of social inequality, especially gender, cannot remain an optional extra. Ignoring gender leads to a reinforcement of the status quo, and even an increase in social and economic inequality, exacerbating stress and social injustice within communities. If the aims of water for all and water as a right are to be met even partly, these issues have to become central in water supply systems (Chancellor, Joshi, Pandey and Moffat).

The Different Needs for Water: The Case of HIV/AIDS

The different roles women and men have in relation to water collection and the way they allocate water use domestically vary between cultures and in relation to factors such as water sources, water availability, religion and geography (Coles, Joshi and Fawcett, Joshi, and Strang). A new event shaping the need for and use of water is the AIDS pandemic in countries of sub-Saharan Africa and now increasingly Asia, yet this has been largely ignored. The impact of HIV/AIDS on the productive members of communities and on changing water demands for treatment and care, is highly significant and will increasingly affect who can participate in water projects and collect water, and how water is to be managed and used. The role of women in the provision of care is adding to their many work burdens, including the need to collect more water, often in contexts of water shortage (Hutchings and Buijs).

Conclusion

Overall the book draws on historical evidence to show that gender relations have determined access to and use of water over centuries. The chapters show that gender relations vary between cultures, even cultures sharing similar geographical spaces, and they illustrate the specificity of how gender beliefs, traditions and attitudes affect water ownership, control of water, water use and the value put on water. Ideologies and images of water have their origins in religion and culture; social constructs of water are strongly gendered. Water providers need to find ways to address complex gender issues, deeply culturally embedded, in the planning, design and implementation of water projects if they hope to achieve the goal of universal access. Experience suggests that while this is acknowledged and discussed at the level of rhetoric, practice varies widely – in reality gender issues are often not seriously addressed.

In bringing together these chapters, which cross so many disciplinary and cultural boundaries – both organizational and geographical – we hope that this book will generate discussion and debate as it has done for us as editors. We also hope that it will stimulate thought, open up the discourses around water and generate ideas that may lead to more effective provision of water for all. Read and enjoy!

References

Aegisson, G. (2002), *The Great Water Robbery*, London: One World Action.

Barlow, M. and T. Clarke (2002), *Blue Gold: The Fight to Stop the Corporate Theft of the World's Water*, New York: The New Press.

Davies, J. and J. Davies (2002), 'The Water Beneath Their Feet', Paper presented to IGS seminar, Queen Elizabeth House, Oxford University. Unpublished.

DFID (2004a), *Consultation Meetings With People From the UK Water Sector on the MDG and a New Policy for DFID*, London: DFID. Unpublished.

—— (2004b), *Accelerating Action on Girls' Education: Rights and Responsibilities*, Draft report on MDGs, London: DFID. Unpublished.

Director WHO (2004) in 'World Facing a Silent Emergency as Billions Struggle Without Clean Water or Basic Sanitation Say WHO and UNICEF' 26 August 2004, http://www.who.int/mediacentre/news/releases/2004/pr58/en.

Elson, D. (1991), 'Male Bias in Macro-Economics: The Case of Structural Adjustment, in D. Elson, ed., *Male Bias in the Development Process*, Manchester: Manchester University Press.

Gender and Water Alliance (2000 onwards), 'Mainstreaming Gender in Water Resources Management', http://www.genderandwateralliance.org.

Ghosh G. (1989), *Country Overview-India. Women and Water*, Proceedings of a Regional Seminar, Manila: Asian development Bank.

Grossman, A., N. Johnson, and G. Sidhu (2004), *Diverting the Flow: A Resource Guide to Gender Rights and Water Privatisation*, New York: Women's Environment and Development Organisation.

Guijt, I. and M.K. Shah (1998), *The Myth of Community: Gender Issues in Participatory Development*, London: ITDG Publications.

Gutierrez, E., B. Calaguas, J. Green, and V. Roaf (2003), *New Rules, New Roles: Does PSP Benefit the Poor?* London: WaterAid and Tearfund on WaterAid website.

INSTRAW (1985), *Involvement of Women in the Choice of Technology and Implementation of Water Supply and Sanitation Projects*, Women and water decade, New York: INSTRAW.

Joshi, D. (2002), 'The Rhetoric and Reality of Gender Issues in the Domestic Water Sector – A Case Study from India', PhD thesis, University of Southampton, Faculty of Engineering and Applied Science.

Kabeer, N. (1994), *Reversed Realities: Gender Hierarchies in Development Thought*, London: Verso.

Kandiyotti, D. (1988), 'Women and Rural Development Policies: The Changing Agenda', *IDS Discussion Paper*, Sussex: IDS.

Levy, C. (1998), 'Institutionalisation of Gender Through Participatory Practice', in I. Guijt and M.K. Shah, eds, *The Myth of Community: Gender Issues in Participatory Development*, London: ITDG Publications.

Longwe, S.H. (1991), 'Gender Awareness: The Missing Element in Third World Development Projects', in T. Wallace and C. March, *Changing Perceptions: Writings on Gender and Development*, Oxford: Oxfam.

Molyneux, M. (1985), 'Mobilisation without Emancipation? Women's Interests, State and Revolution in Nicaragua', *Feminist Studies*, 11(2).

Moser, C. (1993), *Gender Planning and Development: Theory, Practice and Training*, London: Routledge.

Mukhopadhya, M. (1984), *Silver Shackles: Women and Development in India*. Oxford: Oxfam.

—— (1995), 'Gender Relations, Development and Culture' *Gender and Development*, 3 (1), Oxfam.

Painter, G. (2004), 'Gender, the MDGs and Women's Human Rights in the Context of the 2005 Review Process', Paper presented to Gender and Development Network, London: Womankind. Unpublished.

Polaris Institute (2004), 'How Corporations Plan to take Control of Local Water', http://www.polarisinstitute.org/pubs/pubs_global_water-grab_intro.html.

Transnational Institute (TNI) (2004), *Reclaiming Public Water: Participatory Alternatives to Privatisation*, Amsterdam: TNI.

UN (1994), Report of the International Conference on Population and Development, Cairo 15–13 September. UN publication E.95.XIII.18.

Walker, B. (1993), *Misunderstanding Gender*, Gender and Development Unit, Oxford: Oxfam. Unpublished.

Wallace, T. and C. March (1991), *Changing Perceptions: Writings on Gender and Development*, Oxford: Oxfam.

Welbourn, A. (1991), *Rapid Rural Appraisal (RRA) and the Analysis of Difference*, RRA notes, 14.

Whitehead, A. (1979), 'Some Preliminary Notes on the Subordination of Women', *Institute of development Studies Bulletin*, 10 (3), Sussex: IDS.

—— (2003), 'Gender and PRSPs'. Paper written for Gender and Development Network, London: Womankind. Unpublished.

WHO/Unicef (2004), *Joint Monitoring Programme for Water Supply and Sanitation*, Geneva: WHO.

World Bank (1993), *Water Resources Management*, a policy paper, Washington, DC: The World Bank.

World Bank Watch (2004), http://www.citizen.org/cmep/water/new/wbwatch/articles.

World Water Council (2000), 'World Water Vision: Making Water Everybody's Business', in W.J. Cosgrave and F.R. Rijsberman, eds, *Making Water Everybody's Business*. London: Earthscan.

World Water Forum (2003), *Gender and Water*, for meetings in Kyoto, Shiga and Osaka, Japan. Unpublished.

2

Taking the Waters: Cosmology, Gender and Material Culture in the Appropriation of Water Resources

Veronica Strang

Introduction

Water has always been a highly contested resource. Under contemporary population and developmental pressures, issues of ownership and control are becoming increasingly controversial. As Wittfogel commented presciently in the 1950s, 'control of water is inevitably control of life and livelihood'. As soon as humans moved away from hunter-gatherer lives, the planned control of water provided the opportunity for 'despotic' patterns of government and society (Ward 1997: 32). Many writers have noted the relationship between water, social agency and political power, but relatively little attention has been given to the issue of gender in human interactions with water. Similarly, research has rarely considered the gendered meanings encoded in water, or how these are manifested in material terms.

Water is invariably encoded with powerful themes of meaning, and these resurface in different forms in every cultural context, exerting a major influence on the contests for the ownership and control of water resources through which many groups – and it seems particularly women – have been dispossessed. The purpose of this chapter is therefore to consider how the social, political and economic appropriation of water resources happens over time, and how it is invariably accompanied and enabled by an interaction between material artefacts, cosmological beliefs, and concepts of nature, culture and gender.

To illustrate these relationships, the chapter charts some distinctive patterns of historical change in a particular ethnographic context, attempting to provide a larger 'bird's eye' view of an evolving cultural landscape. The material is drawn from a three-year research project based in the Stour River Valley in Dorset, in which the collection of ethnographic data and oral histories was accompanied by the extensive use of archaeological evidence, local archives and literature, maps and other forms of representation. A detailed account of contemporary issues in the area is offered elsewhere (Strang 2004).

Distilling History

Contemporary conflicts are deeply rooted in the past, and a historical overview reveals how these have been shaped into particular patterns over time. In Dorset, narratives of local water ownership and use – and the first appropriations of water – flow a long way back. Archaeological evidence dates some activities from approximately 5000 BC, and there are some useful (albeit sketchy) accounts in the reports made by early Roman colonists. According to Ptolemy, Dorset was originally inhabited by Celtic tribes called Durotriges, meaning 'water dwellers', who hunted and gathered in the forests that dominated the landscape, reportedly 'devoting themselves to the superstitious rites of their religion' (Hutchins 1973 [1861]: ii).

Though little evidence remains, it is reasonable to surmise from more recent studies of hunter-gatherers that the Durotriges' socio-economic organization was likely to be have been based on kinship and exchange, and on gerontocratic – and thus egalitarian – forms of political governance. There was probably some gender role division in terms of hunting and gathering, and we can also assume uncontroversially (as this is a basic necessity for hunter-gatherers) that the Durotriges had intimate knowledge of their local environment. With a mode of production requiring regular patterns of movement around a particular landscape, their material culture was minimal: simple water vessels, stone tools, leather and woven materials, and later some iron or bronze artefacts and ornaments.

Archaeological findings and Roman descriptions of their religious practices suggest that they had numerous deities, and that both female and male gods were located in their immediate environment, in a sanctified landscape. In accord with Durkheimian theories concerning the reflective relationship between cosmological beliefs and socio-political organization, this implies considerable political, social and economic equality for women, and collective ownership and control of resources.

There are detailed accounts of water cults in Egypt, classical Greece, Troy, Babylon and Rome, and considerable evidence that water worship was practised by Celtic tribes right across Europe. Water symbols have been found on images of goddesses and other figurines dating from 6000–4000 BC (Bord and Bord 1985), and existing river names are frequently derived from the names of Celtic goddesses.[1] Some rivers, such as the Seine (named for the goddess Sequana), had Celtic temples at their sources, and there are persistent relationships between religious material culture and water sources or rivers. For example, Richards (1996: 313) considers water as a major component in the physical and symbolic arrangement of Neolithic henges, and at Dorchester's Maiden Castle – one of the greatest Iron Age forts in Britain – Hawkes (1951) notes that there is a stone enclosure with a rough image of the Mother Goddess.

Early Celtic religious practices apparently centred mainly on wells and trees. The Roman poet Lucan (AD 39–65) described a forest sanctuary: 'there were many dark springs running there, and grim-faced figures of gods uncouthly hewn by the axe from the untrimmed tree-trunk' (Bord and Bord 1985: 7). The historian Livy (59 BC–AD 17) noted that Celtic warriors used skulls decorated with gold to make offerings to the gods at holy wells, and many other votive offerings have been found in these. For example Caesar wrote about the finding, in a well near Dorchester, of a bronze offering stamped with the sign of a hare – an animal apparently sacred to the Celts. In general terms, the material evidence suggests a gendered homologous relationship between wells and trees, henges and water, with standing stones and trees providing a male counterpart to the containing womb of the circle and the well.[2] The implication is that Celtic religion, like that of other hunter-gatherers, was characterized by gender complementarity, and that water was firmly associated with feminine principles.

Occupational Hazards

In Dorset, the Durotriges' way of life was supplanted by increasingly forceful invasions. First, Celtic tribes from Brittany moved up the Stour to Marnhull, bringing with them animal husbandry (oxen, pigs and horses), agriculture (wheat and celtic beans), and trade. Like agricultural innovations and introductions elsewhere, this widened gender roles, and encouraged new – and less egalitarian – forms of property ownership.

The Romans invaded Britain in 55 BC, and by AD 43 the whole island had been 'reduced to obedience' and the inhabitants enslaved. This was primarily a military occupation, with a specific remit to report back: *veni, vidi vici*. This placed the Romans in a stance that was adversarial both to the local inhabitants and to their environment, and there is plenty of archaeological evidence of their success in conquering both.

Roman society was more complex than that of the Celtic tribes, and the male-led hegemonic military enterprise brought rapid technological advancement, enabling the invaders to replace a relatively egalitarian society with a steeply hierarchical one in which there were major class and gender divisions. Their social impositions also required religious dominance: in AD 45, for example, they launched a full-scale assault on Maiden Castle and, having successfully overcome Celtic resistance, built a Roman temple inside its ramparts, asserting the primacy of their own beliefs and practices.

Nevertheless, the Romans retained many cosmological concepts suggestive of an earlier, more intimate interaction with a supernatural landscape. In conquering Britain, they either appropriated or imported numerous water goddesses. Like the

Celts, they also made votive offerings at wells and streams, and their annual ritual, Fontanalia, involved elaborate well dressings. As well as focusing on water in their religious rituals, they made extensive use of baths and spas, which they believed to be essential for good health. For example, in Bath, a hot spring identified with the Celtic goddess of healing, Sulis, became associated with the Roman healing deity, Minerva (Irvine 1989a).

Roman cosmology contained numerous female and male deities primarily concerned with healing and fertility. Dorset is particularly well endowed with these, being the home of the famous Cerne Abbas Giant a few miles north of Dorchester. Perhaps less well known is the holy well at the base of the hill upon which the giant lies: a juxtaposition that suggests gender complementarity in the meanings encoded in the landscape. However, it is also possible to interpret the addition of the looming Herculean giant (which some archaeologists believe subsumed a previous figure) as a shift in gender balance in the Roman cosmos.

The Roman domination of Britain was both religious and material. The colonists went to great lengths to bring water under close control, using their advanced technological skills to channel it through aqueducts, baths and spas. They also built one of the first water mills in Britain. In this way, although their pots and mosaics suggest that women retained a central role as the 'water carriers', both literally and symbolically, the wider management of water resources began to shift into the hands of male engineers.

Waves of Invasion

The male appropriation of water was greatly enabled by the next invasion, which was largely ideological. The early Christianity that spread through Europe was intensely intolerant of 'Pagan' forms of worship. In 452, a religious canon decreed that any bishop permitting 'infidels' to venerate trees, fountains or stones, would be considered guilty of sacrilege (Bord and Bord 1985). Water worship was immensely persistent though, and in the sixth century, recognising this, Pope Gregory recommended that, rather than being destroyed, temples and holy wells should be converted for Christian use. Thousands of holy wells throughout Britain and Ireland[3] were thus subjected to saintly takeovers and makeovers. The holy well at the foot of the Cerne Abbas Giant was renamed St Augustine's, and the story of this conversion suggests that, while asserting male dominion over the well, St Augustine also made a wonderfully phallic challenge to the Herculean figure looming on the hill above. According to Thomas Gerard, writing in the 1620s, he 'pitched his Staffe on the Ground ... when immediately flowed a quick Fountaine, that served to baptise manie' (Irvine 1989b: 40).

Thus the homologous image of the female well was superseded by St Augustine's spouting 'staffe', reframing water as a product of human (male) creativity. This story therefore encapsulates a crucial shift in cosmology: a separation of culture from nature, that placed them, for the first time, in opposition to each other. The Judaeo-Christian beliefs that flowed into Dorset contained concepts in which God or spiritual being was perceptually removed from the landscape. A more abstract and distant 'God the Father' came to represent the civilized antithesis to uncultured nature. In this way land and water became both the creation and subject of God's – and thus *man*kind's – will, validating the patriarchal control of resources and the subjugation of women.

In Shaftesbury, near the head of the Stour, King Canute passed laws in 1018 prohibiting the worship of 'heathen gods', including 'rivers, fountains, rocks and trees' (Irvine 1989a: 32). However, Canute remains most famous for his humiliating failure to 'stem the tide', so it seems that water power continued to flow, regardless. Such decrees continued to emerge from the Church into the twelfth century, yet Christian imagery retained vestiges of an older cosmos characterized by gender complementarity. Many holy wells (and their still resident, if silenced, goddesses) were encrypted with sacred fish, a classic phallic fertility symbol. Conflicts between Christian (male) power and more ancient Pagan (female and male) forces continued to be played out at religious and political levels.

By the latter end of the first millennium, the Roman era had passed and Saxons and Danes were vying for control of Dorset. The symbolic power of water in these contests is nicely illustrated in local legends. For example, the six springs that are the source of the Stour are said to have risen in response to King Alfred's prayers for water for his men, in the war with the Danes (Hutchins 1973 [1861]: 97). Another legend is located at Halstock, which means 'holy place'. Here, at a spring dedicated to the Saxon saint Judith (or Juthwara), Irvine records that: 'She is reputed to have been beheaded by her brother, a Pagan, when she became a Christian. Undismayed by the event she picked up her head and carried it to the church where she placed it on the altar. At the spot where her blood was spilled, a spring of clear water welled up from the ground' (1989b: 38).

This is an interesting parable: on the one hand it can be interpreted as presenting a heroine, under attack by 'barbaric' Paganism, choosing to bring feminine spiritual being into the Church. The beheading, and the placing of the head upon the altar, suggest the offering of the housing of the intellect to a 'rational' Christian God. However, the welling up of spring water from the site on which Judith is killed, outside the Church, strikes a slightly subversive note, suggesting the continued flow of resistant female spirituality.

Like the Romans, the Saxons retained a strong undercurrent of Pagan beliefs in the spiritual power of water. They used water from sacred springs in charms and ritual practices, and believed that drawing water 'at sunrise and in silence from an eastward flowing stream' would 'restore youth, heal eruptions and make cattle strong' (Irvine 1989a: 31–32). Some wells were named for female saints, and many were regarded as 'healing' wells right into the eighteenth century, being commonly used to cure ills and assist fertility. However, over time most holy wells were enclosed, and the water channelled so that it would spout from a Christian 'font'. In Dorset, a chapel was built over St Augustine's well in the Middle Ages, and it became an important centre for pilgrimage (Irvine 1989b: 40).

There were similar advances in the material control of the landscape. The Saxons were highly organized farmers who totally reshaped the area, which they called Dornsaeta. They established local and central governance in the form of sheriffs (shire-reeves), who acted for the King in managing shires, hundreds, hides and other subdivisions of land and resources. As Bettey says: 'By the year 1000 ... Saxon pioneers had laid out the landscape with a framework of estates and villages, towns and parishes, administrative boundaries and ecclesiastical divisions' (1986: 5). The intensification in farming – and the population increases it permitted – resulted in the first major land enclosures in Dorset. The loss of collective resource ownership gathered momentum as land and water rights came to be held by hereditary male lineages. By the Norman conquest in 1066, most of the land in Dorset was already part of large manorial holdings. Some areas were taken over by the King as Royal Forest, and large estates were owned by powerful abbeys such as Glastonbury and Shaftesbury. Like the manors, the abbeys often took over control of water resources, considering it part of their religious duty to dispense water to their dependent tenants.

Material Dispossessions

The use of more sophisticated farming technology created greater separation in gender roles, though women were still the primary water carriers. They gathered daily at village wells, to collect water and exchange information, and so maintained a powerful female political and social space. Gradually, village wells were modernised, and there was another homologous transition in material culture as holding wells were replaced by extractive pumps. As Bailey comments, these village pumps remained central, protecting common water supplies from vandals or pollution: 'They stand like symbols of village continuity, and ... the community's source of life' (1985: 22–23).

In the wider landscape, water resources were also subjected to increasing technological control. The Domesday Book (1086) records 166 water mills in Dorset, usually milling corn or grain. Every stretch of river was thus carefully controlled through a system of checking and releasing water flows. Over the next several centuries, the fertility of the Dorset river valleys created a class of yeoman farmers wealthy enough to buy their own land, thus subdividing land ownership further. In the 1600s there arose a system of water management making extensive use of meadows, in which a system of sluices and hatches kept water on the land in the winter. As Bettey explains: 'The object of a water-meadow was to cover the grass with a thin blanket of running water from the chalk stream, thus maintaining it at a steady temperature, protecting it from frost, and depositing valuable silt and sediment around its roots' (1986: 26).

By hastening new growth, water meadows enabled farmers to feed sheep or cattle between the end of winter and spring growth, enlarging the carrying capacity of the land.[4] Many thousands of acres beside the river were refashioned in this way. The building of such an elaborate system of water management required engineering or carpentry skills, and was managed by 'watermen'. Similar advances were made in the general management of water sources, where pumps and pipes were installed to improve the control of supplies.

Technological and scientific change burgeoned, and in the age of the Enlightenment the seeds of rationalism sown in early Christianity grew to fruition. The landscape-based folk tales and early fertility symbolism of the Old Testament were superseded by ideas about the need for rational male 'dominion' of the earth. In the New Testament, rather than focusing on wells, images of water reflected the idea that it was the product of God. Like St Augustine's spouting 'staffe', the fountainhead was represented as an active, living spring flowing from God. As Schama points out, the fluvial literature of the fifteenth and sixteenth centuries was obsessed with images of the fountainhead as the *fons sapientiae*: the mystically revealed union of goodness, beauty and wisdom (1996: 267). The Enlightenment was seen as a final stage in a process of intellectual abandonment of foolish superstitions and the beliefs of the 'dark ages', opening the door to science and a material, Cartesian view of the environment.

These beliefs were materialized at Stourhead in the 1730s, when Henry Hoare, imitating Italian Renaissance architecture, built his famous landscaped gardens. The 'six wells' that had risen miraculously to assist King Alfred, were enclosed by St Peter's Pump and channelled to emerge in a 'Grotto'. With the spirit of nature looking on (in the form of a statue of a nymph), the source of the Stour emerged as an 'enlightening' fountainhead, flowing into an intensely cultured landscape with a carefully constructed lake, and a spiritual 'journey' linking a series of classical temples with St Peter's church.

The intensification of farming and production was therefore matched by attempts to mould an aesthetic landscape in accord with particular cosmological precepts. Though still working on the land as crop planters and gatherers, women were not part of this directive managerial activity, although they were apparently able to be 'at one with nature'. The framing of gender roles during this period is nicely illustrated in Thomas Hardy's novels of Dorset life. In *Tess of the D'Urbervilles* – located in Marnhull, in the upper reaches of the Stour – Hardy refers to 'the charm which is acquired by woman when she becomes part and parcel of outdoor nature, and is not merely an object set down therein... A field-man is a personality afield; a field-woman is a portion of the field; she has somehow lost her own margin, imbibed the essence of her surrounding, and assimilated herself with it' (1891: 87–8).

Subversive Streams

Despite all the efforts to mould the landscape to express Christian morality and rational philosophy, other ideas continued to bubble along subversively under the surface. Pennant, touring the country in 1769, found that holy wells were still much visited, with coins and rags being left at them as offerings. As he put it:

> Holy wells are probably far from the thoughts of persons living amid the stir and bustle of city life, but in rural districts, where old customs linger, they are not yet forgotten. In the country, amidst the sights and sounds of nature, men are prone to cherish the beliefs and ways of their forefathers. Practices born in days of darkness thus live on into an era of greater enlightenment. (Mackinlay 1993 [1893]: 14)

However, he added that: 'though modern enlightenment has not entirely abolished the practice of resorting to consecrated springs, it has, as a rule, produced a desire for secrecy on the part of the pilgrims' (Mackinlay 1993 [1893]: 278).

In the course of this enlightenment, 'nature' was deconstructed, demystified, and opened up for analysis. Concepts such as germs, bacteria and infection replaced vague ideas about 'miasmas' and mysterious energies, and water became H_2O. For the educated classes, its healing properties acquired 'scientific' explanations. 'In the seventeenth and eighteenth centuries water had a new lease of life as a healing medium when mineral springs and spas became fashionable. While the holy wells appealed to the simple country folk, the spas were patronised by the gentry' (Bord and Bord 1985: 52).

Nevertheless, even in the 1880s, Moule recorded that local people 'hold to the belief that St Austin's Well ... still works wondrous cures' (Irvine 1989b: 40). As Irvine observes, the water had retained its reputation as a cure for infertility among women:

> a Victorian journalist, writing in a local paper, made a discreet allusion to this property of the water... 'We should recommend lady travellers passing through to rest a while and pay a visit (to the well) and they will experience ... the happiness of having their wish granted within a year.' If a woman did give birth to a child, it was considered beneficial to dip the new-born infant into the waters of the spring 'at the time when the sun is first shining on the water' ... this reflects a Saxon custom practised over a thousand years before. (1989b: 40)

However, Mackinlay notes that: 'agricultural improvements, particularly within the present century, have done much to abolish the adoration of wells. In many cases, ancient springs have ceased to exist, through draining operations' (Mackinlay 1993 [1893]: 17–18). New forms of 'scientific agriculture' required large amounts of capital and, as Roscoe says:

> the landlords and their tenants profited very greatly while the class of independent yeoman farmers, once so numerous, was almost extinguished. The old communal system was incompatible with the new style of farming. So, for more than a century from about 1740, enclosures were everywhere pressed forward. The effects on productivity were remarkable, but … the humbler peasants, paid the price. Deprived of their immemorial rights in the land they were either driven to the towns or reduced to the depressed condition of mere farm labourers. (1952: 49)

The Napoleonic wars required even larger scale production, encouraging more enclosures and more centralized forms of governance. The beginnings of the Industrial Revolution also pushed many farmers off the land and prevented public access to former commons. Local archives in Dorset record widespread protests against the enclosures: the burning of agricultural machinery and buildings, and such violent riots that there were calls for the military to be brought in. Resistance proved futile though, and most of the population – not just women – lost their remaining control over land and water resources.[5]

These developments, and the Industrial Revolution as a whole – which was largely driven by water power – made water resources of primary importance. Large infrastructural developments, such as canals and reservoirs, required government regulation in the management of water, which further reduced local resource control. The demographic shift into the cities also had huge implications for water management, creating large centres of demand. At first urban settlements relied upon local public wells and pumps, and the piping in of 'sweet' water from further afield was a luxury for the wealthy few. Soon though, industrial activities such as tanning, and the heavy concentration of people (and therefore sewage effluent) in small areas, meant that rivers in urban centres became so polluted that typhoid and cholera epidemics began to occur with alarming frequency.

Taking the Waters

Recognition of the need for clean drinking water in urban areas led to the establishment of the first private water companies, set up with shares to provide the capital for infrastructural investments. One of the earliest was in Weymouth, where a 1797 parliamentary bill licensed the company to supply 'the Borough and Town of Weymouth and Melcombe Regis, and the Parts adjacent, in the County of Dorset, with Water'. It noted that: 'the Borough and Town of Weymouth and Melcombe Regis, and the Parts adjacent, which are very populous, are not properly supplied with fresh Water ... it is of great Consequence that the Inhabitants thereof, and the Shipping resorting thereto, should have a constant Supply' (Dorset County Archives). Other private and municipal water companies sprung up elsewhere. Water sources were therefore enclosed either by local authorities, or by privatized water companies, both of which were led by male directors, supported by the expertise of male water engineers and chemists.

Though essential for public health, this put an end to collective water resource control in urban areas. The 'village well' disappeared from within the cities, and – in an era of curtailed female property rights – women lost not only the labour of carrying water, but also any remaining part in the ownership, control or management of water resources, becoming largely passive recipients of water: the 'consumers' of private or public supplies.

As 'nature' was domesticated, so too were the images of the relationship between women and water. Illich notes how painters of the nineteenth century began to present the female nude as a bather, which 'tied female nudity as a cultural symbol to the tap water of the bathroom... The proximity of suds and nude in the bath domesticated both water and flesh' (1986: 1). At the time, piped domestic water was seen as quite a benevolent form of control over water. It represented a luxurious saving of labour, and the urban need for clean water and complex infrastructure left little choice.[6] Although control of water resources had passed upwards to distinctly patriarchal institutions, most water companies remained publicly owned. With vestiges of the paternal responsibilities exercised earlier by local abbeys and manors, the supply of water was seen by many as a moral and semi-religious duty, a responsibility of the city 'fathers'. New supply or waste systems were frequently initiated by church groups and wealthy philanthropists, and as Ward points out: 'This binding moral duty to ensure that every household had access to a clean water supply and safe disposal of wastes, was slowly accepted everywhere... Hence the great monuments of Victorian water and sanitary engineering that astonish us today' (1997: 6–7).

Nevertheless, there was an ongoing tug-of-war between public and private ownership, and a growing tension between concepts of social responsibility and commercial interests. Private companies tended to 'cherry pick' profitable areas, and. as Ward adds, 'in the latter half of the 19th century, faced with the fact that private water companies had no interest in providing water supply to poor households, local authorities were obliged to buy the water companies to make provision universal' (1997: 96). In 1885, over 100 years before the Thatcherite privatization of 1989, commercial water suppliers banded together in an Association to oppose government regulation, arguing – as the industry does today – that its expertise justified its private control of water resources.[7] However, many local authorities stepped in and, bolstered by numerous public health acts and sanitary acts, the number of public water suppliers proliferated. By 1910 there were 2,160 in England and Wales, of which only 152 were private statutory companies.

In Dorset, as elsewhere, the population and the number of households rose rapidly, placing considerable pressure on water resources. As the scale of resource management increased, water companies merged and enlarged, halving in number by 1925. The government and its regulatory bodies underwent a similar pattern of centralization, forming larger and larger institutions. However, the Victorian ethos about universal access to water persisted. In a post-Second World War era characterized by the political conviction that wealth should be distributed more fairly, water resources were returned, via the 1945 Water Act, to democratic, collective ownership. The process of centralization continued: in 1963 the water supply and sewage companies were amalgamated to create a mere twenty-seven river authorities, and in 1974 this number fell to ten regional water authorities.

Although these mergers enabled some efficiencies, water supply technology demanded major injections of capital, and there was a reluctance to risk the political costs of making sufficient public investment. The magnificent Victorian infrastructure was crumbling, and there were increasing environmental concerns and legislative demands to reduce pollution and improve water quality. In 1989, despite massive public protests, the Conservatives privatized the entire sector, creating a virtual monopoly. Although this was supposed to be ameliorated by government regulation, water prices rose by about 60 per cent over the next eight years, were held in check briefly by the Labour government's first price review,[8] and continued to rise subsequently. Profitable share deals and other commercial advantages encouraged a rash of foreign takeovers, to the extent that over 40 per cent of water companies in the UK are now owned by foreign corporations.

Since the privatization of the industry, largely removed from local democratic structures,[9] centralized government regulatory bodies and the industry have wrangled over prices and profits, and technical issues, rather than more fundamental questions about access and ownership. Ideologically they appear to be in semi-agreement about the reformulation of water as a commodity, although this balance shifted somewhat when the new Labour government was elected.

Contemporary Cultural Landscapes

What have these changes meant for water users along the Stour? At the beginning of the twenty-first century, the Stour Valley is fairly typical of many parts of the UK, although demographically it has a slightly older and wealthier than average population. Housing in upstream rural villages is heavily dominated by retirees from London, while the younger generations and most of the employment are located in the downriver urban sprawls of Bournemouth, Christchurch and Poole. Almost all of the land is privately owned, and it is used with enormous intensity, creating a highly manicured, heavily regulated and boundaried cultural and physical landscape.

The number of households in Dorset has almost quadrupled over the last century, and over 98 per cent now have piped water and modern bathing facilities. For most people this domestic usage is now their only access to water resources, unless they undertake recreational activities – fishing, boating and suchlike – which take place in carefully circumscribed areas. Most river banks – and indeed most of the land areas – are accessed only through a few remaining rights of way. Small recreational or conservation groups offer some input to local water resource management, but this is now primarily the province of the major landowners, privatized water companies and the centralized government agencies such as the Environment Agency and – more distantly – the Drinking Water Inspectorate and OFWAT.

People's sense of alienation from involvement in resources emerged strongly in the ethnography. Opposition to the privatization of water resources remains passionate: as one informant put it, echoing commonly expressed sentiments, 'it was the worst thing they ever did to us'. There is deep antipathy towards the major regional company, Wessex Water, most particularly since its takeover by the American consortium Enron in the later 1990s. As one informant put it: 'I think it's appalling really ... I feel very upset about ... we are losing control over our resource' (David Solly).

There are significant continuities in the powerful meanings that people encode in water (see Strang 2004). Many retain ideas about water as a healing substance, imbued with particular energies: 'Water has always had this healing hasn't it' (Beryl Coward), or 'it is calming and healing' (Edward Jacson). There is still some informal ritual activity at wells and springs, and 'neo-Pagan' groups have returned to ideas about the well as 'the eye of the Earth Mother', or entry to 'the womb of the Mother'. A woman living in an 'alternative' community described visiting an ancient well to perform a Pagan baptism:

> I did baptise my oldest [child] myself in water ... I go along with the ley-line theory that there are special points of high energy and that often there are wells in these points, and therefore the water is charged with more energy than other sources of water ... those older sort of ideas ... people have just carried on with the same idea but maybe the religion has changed. (Sue Degan)

However, Christian and scientific ideologies remain dominant. In the former, 'the fountainhead' of the Enlightenment is still the central image; thus a vicar describes baptism as 'being filled with the spirit … I just accept the images, what the Bible says … one very important image is this stream running from the temple, from the throne' (Colin Marsh). In scientific terms, hydrology has subsumed hydrolatry: seen as integral to a material ecology, H_2O is measured and read in microscopic detail. These ideas are not incompatible with religious views: in a sense, Cartesian concepts fit quite comfortably inside the larger abstractions of Christianity, offering a material world that is the product of God or, in Durkheimian terms, the product and subject of (male) culture.

Conclusion

A number of important patterns can be traced in the historical narrative. Like other depictions of modernity (for example, Berger and Kellner 1973), these reveal exponential increases in scale and suggest a general process of unravelling, fragmentation and divergence, leading – ultimately – to a sense of alienation. In cosmological terms, highly localised land-based religions containing powerful – and equal – male and female deities became, over time, divided by gender. Their feminine principles were demoted via a much more abstract concept of nature, and made subject to culture as a rational, controlling force. The concurrent political creation of complex hierarchies dispossessed women of economic and social parity. Through a series of enclosures and privatizations, the majority of the population, but women in particular, were steadily alienated from collective or direct ownership or management of resources, with control or the expertise to manage held by an ever smaller minority. Local involvement with resources vanished in favour of centralized forms of governance and resource management, so that contact with water became almost totally located in an individuated domestic environment.

All of these separations have been concretized and enabled by related material culture which has echoed, in its form, the gender issues implicit in this historical overview. Thus the material culture of water supply demonstrates a progression away from archetypal 'female' objects, such as wells and containing vessels, to more phallic material culture: St Gregory's 'staffe'; gushing pumps and taps; and that ultimate statement of powerful male agency and social status, the spurting fountain. Such potent status symbols are the province of a lucky few, although individual efforts to emulate and thus share this power are revealed in the increasing popularity of water features in gardens (see Strang 2001 and 2004).

Technological change has also enabled the physical alienation of water not only from women, but from local communities and, eventually, from the bulk of the population. The social and religious locus of water use moved from the forest spring to central village wells, and then to pumps on the village green. These focal points were replaced, first by piped and channelled water from local sources, then by water from further afield, and, over time, by a vast underground infrastructure capable of carrying water for many hundreds of miles. With increasing technical complexity, this material culture became more and more exclusively the province of experts, so that contemporary water resources are controlled largely by engineers, chemists and computers. Thus the physical management of water that used to be everyone's business, and especially women's, is now carried out by a very tiny number of people, the vast majority of whom are men.[10]

Much of the material culture that now exerts so much control over water is invisible or removed from the domestic sphere. Large wells and pumps abstract water from aquifers; a series of sophisticated filters and chemical processes turn it into the required potable substance; a huge network of mains and smaller pipes sends it under pressure to individual homes. Here it emerges in domestic kitchens and bathrooms as if from a spring under the house, courtesy of a regional water supply company. The new material culture – individual supplies to single dwellings, household meters and so on – coheres with a reality in which, increasingly, water companies have attempted to reframe water as a commodity: a product of their labours. With meters, water is paid for in measured units to which a price can be more readily attached. Water users – by executive edict – are now resolutely described as 'consumers' or 'customers'.

Through this combination of ideological and material change, ordinary water users have become the passive recipients of water supplies owned and controlled by a small elite of shareholders, managers, regulators and experts, under the aegis of a centralized government, and vast international companies whose social relationships to the communities they purport to serve are at best tenuous.

The environmental and social consequences of centuries of alienation and commodification have been considerable. Many aquatic ecosystems and vital habitats have been lost with forest clearance, the draining of wetlands, intensive farming, and industrial development and pollution. Despite the increasing frequency of winter floods (caused in part by intensive development), over-abstraction from aquifers now causes some rivers to run dry in the summer. Many informants in Dorset commented that the Stour contains much less water than previously. Other rivers around Dorchester, such as the presciently named River Piddle, have at times almost disappeared, with commensurate losses in aquatic species.

There are several major causes of over-abstraction. Intensified farming, as well as requiring high levels of irrigation, has led to the removal of the storage 'sponges' provided by forests and wetlands. With drinking water supply, it is safer and cheaper (and therefore more profitable) to abstract clean water from upriver sources than to treat polluted downriver water. Crucially, along with increasing population density, levels of domestic water usage have climbed steadily since the first introduction of piped supplies, and per capita use has more than doubled since 1961 (Water Services Association, 1997). The complex reasons for this dramatic rise in demand are addressed elsewhere (Strang 2004), but it is clear that the material culture has reflected and enabled major changes in practice. As well as being economically and physically alienated from water resource management and control, domestic users' experience of water supply provides no readily discernible relationship between the gushing individual spring of a household tap and the source of the water. Many people don't even know where their household water comes from, relying instead on a vague impression of being tapped into a vast system of infinite supplies. Such detachment from visible ecological processes has enabled users to forget that water resources are actually finite, and in a country which seems to have ample rain, it has been all too easy to lose touch with the ecological costs of this assumption.

As well as permitting a perceptual separation between consumption and environmental consequences, the reframing of water as a material product or commodity also undermines ideas about social responsibility and equality in access to resources. Intractable opposition to water privatization vies with the stubborn determination of a political and economic elite to maintain it, reflecting a major tension between ideologies in which the 'common good' and collective rights are set against much more individuated and competitive visions. The opposing sides of this divide are by no means balanced in terms of gender politics. Many of the groups most vehemently opposed to the privatization of water have their roots in the environmental movement, which – though this is often forgotten – emerged in tandem with the feminist movement of the 1960s. Concerns for the environment are founded on ideas about balancing human and environmental needs, and the concept of an equitable balance between nature and culture is symbolically linked to discourses about gender equality.

So is there any hope for the Goddess in the well? The historical ethnography demonstrates that she has been steadily disempowered by successive patriarchal institutions, assisted by technological changes that have asserted the dominance of culture over nature. However, the battle for equality is far from over: historical precedent permeates modern conflicts over water, carrying images and material culture reflecting more egalitarian beliefs and practices. All along, the most common users of the holy wells have been women, and today they are still the major actors in modern well dressings and other such ritual performances.[11] Throughout centuries of change, subversive movements have continued to flow under the surface of dominant ideologies. Years of covert Paganism have been followed by contemporary revivals of neo-Paganism, New Age concepts, feminism and environmentalism. Such movements demonstrate the extraordinary persistence of alternative religious beliefs that assert the feminine aspects of nature. Women led the way towards the environmentalism which flowered in the 1960s and

1970s, and feminist environmental activists such as Arundita Roy are regularly embroiled in water issues, such as the massive enclosures represented by dams in India, Turkey and China. Indeed, feminists have always been particularly closely involved in debates about water.

Of course women have not been alone in opposing the appropriation of resources: widespread resistance has met every enclosure and privatization of land and water. In a contemporary political context, dissent about the ownership of resources emerges in direct action: people chain themselves to trees, fight for the 'right to roam', and refuse to pay water bills on the basis that they have a right to water. At a broader political level, concern about social and environmental equity emerges in discourses about access to water from consumer organizations, environmental groups and other NGOs, who argue for 'universal rights to water' for all, or for the rights of 'nature' not to be sacrificed to profits. Indigenous communities have contributed different cultural perspectives that challenge Western science and industrialized countries' ideas about human–environmental relationships.

These arguments have permeated many of the negotiations that surround water ownership and control, enlivening legislative debates, and applying pressure to the Environment Agency, OFWAT and other agencies that hold regulatory and managerial responsibilities for water resources. Thus the powerful ideologies and the forms of material culture that have placed the control of vital resources into just a few white, male hands are being questioned, and 'alternative' ways of interacting with a social and physical environment – ways that promulgate what are generally framed as 'feminine' principles – continue to challenge this appropriation.

Notes

1. There is the Marne, which comes from Matrona 'Divine Mother'; the Dee, from Deva, 'the holy one' or 'the goddess'; the Clyde, from Clota 'Divine Washer'; the Braint and Brent Rivers, which celebrate Brigantia or Brigit; and the Severne, from Sabrina.
2. Homologues may be described as strings of conceptual associations through which human beings 'project themselves' onto the world by describing it in terms of the human body. They – and their gendered meanings – are manifested in the form of material culture (see Strang 1999).
3. Bord and Bord (1985) calculate that there were at least 8,000 holy wells in Britain and Ireland. Morris and Morris (1981), list over 1,000 in Scotland alone.
4. 'Carrying capacity' refers to the numbers of stock that the land can carry per hectare or acre. In 1859 there were two commons in the parish of Marnhull, of 80 and 25 acres

respectively, but by 1862 these had been enclosed. Over the 1700s and 1800s, twenty separate acts of parliament were passed to clear and enclose woodland in the Vale of Blackmoor alone, and many fens and marshlands were reclaimed and enclosed (Dorset County Archives).

5. Dorset is the home of the Tolpuddle Martyrs, famously transported to the colonies in 1834.

6. These changes came more slowly to rural areas, where the population was able to rely on cleaner local water sources. Even in 1951 nearly a quarter of the households in Dorset still lacked an exclusive piped water supply (HMSO 1955).

7. The Provincial Water Companies Association formed in 1885 became the Water Companies Association. Efforts to assert its expertise were challenged in 1911, when public suppliers formed the Municipal Waterworks Association. Governments set up research units of their own to consider hydrological matters, for example the Water Research Association in 1955. Today the major representative body of the industry is called Water UK.

8. Price reviews take place every five years, and OFWAT's role in these is to regulate the prices water companies are permitted to charge water users for supplies.

9. OFWAT (Office of Water Services) Regional Customer Service Committees, now called 'Watervoice', are intended to provide some degree of public representation, but these committees are unelected, and focus primarily on complaint handling, rather than the larger issues.

10. In the last decade a few female engineers and middle managers have made it into the industry, but the reality that most senior and technical positions are occupied by men is amply illustrated in the annual reports and other publications of the water companies and regulatory agencies.

11. Hundreds of well dressing rituals are now conducted annually in the UK, and these have largely been initiated by women. Most are designed to celebrate local identity and encourage involvement in community life.

References

Bailey, B. (1985), *The English Village Green*, London: Robert Hale.

Barty-King, H. (1992), *Water. The Book: An Illustrated History of Water Supply and Wastewater in the United Kingdom*, London: Quiller Press.

Berger, P., B. Berger and H. Kellner (1973), *The Homeless Mind: Modernization and Consciousness*, New York: Random House.

Bettey, J. (1977), 'The Development of Water Meadows in Dorset During the Seventeenth Century', *Agricultural History Review*, 25: 37–43.

—— (1986), *Wessex From AD 1000*, London: Longman.

Bord, J. and C. Bord (1985), *Sacred Waters: Holy Wells and Water Lore in Britain and Ireland*, London: Granada.

Dewar, H. (1960), 'The Windmills, Watermills and Horse-Mills of Dorset', Reprinted from the *Proceedings of the Dorset Natural History and Archaeological Society*, 82.

Goodland, N. (1970), 'Farming the Water Meadows', *Country Life*, Dorset County Archives.

HMSO (1955), 1951 Census England and Wales. County Report. Dorset, London: HMSO.

Hardy, T. (2003 [1891]), *Tess of the D'Urbervilles*. London: Penguin.

Hawkes, J. (1951), *A Guide to the Prehistoric and Roman Monuments in England and Wales*, London: Chatto and Windus.

Hutchins, J. (1973 [1861]), *The History and Antiquities of the County of Dorset*, vol. 1, Westminster: EP Publishing Ltd and Dorset County Library.

Illich, I. (1986), *H₂O and the Waters of Forgetfulness*, London: Marion Boyars.

Irvine, P. (1989a), 'The Goddess at the Well', *Dorset: The County Magazine*, April: 30–32.

—— (1989b), 'Saints and Spas', *Dorset: The County Magazine*, May: 38–42.

Knight, P. (1998), *Sacred Dorset: On the Path of the Dragon*, Chievely, Berks: Capell Bann Publishing.

Mackinlay, J. (1993 [1893]), *Folklore of Scottish Lochs and Springs*, Glasgow: William Hodge and Co.

Morris R. and F. Morris (1981), *Scottish Healing Wells*, Sandy: The Alethea Press.

Rattue, J. (1995), *The Living Stream: Holy Wells in Historical Context*, Woodbridge, UK: The Boydell Press.

Richards, C. (1996), 'Henges and Water: Towards an Elemental Understanding of Monumentality and Landscape in Late Neolithic Britain', *Journal of Material Culture*, 1(3), November: 313–35.

Roscoe, E. (ed.) and the Marnhull Festival of Britain Committee (1952), *The Marnh'll Book: Some Particular History, Some General Topography, a Number of Photographs and Some Maps of the Blackmore Vales*, Gillingham, Dorset: The Blackmore Press.

Schama, S. (1996), *Landscape and Memory*, London: Fontana Press.

Smith, K. (1972), *Water in Britain: A Study in Applied Hydrology and Resource Geography*, London: Macmillan.

Strang, V. (1999), 'Familiar Forms: Homologues, Culture and Gender in Northern Australia', *Journal of the Royal Anthropological Society*, 5(1), March: 75–95.

—— (2001), *Evaluating Water: Cultural Beliefs and Values About Water Quality, Use and Conservation*, October, London: Water UK.

—— (2004), *The Meaning of Water*, Oxford: Berg.

Ward, Colin (1997), *Reflected in Water: A Crisis in Social Responsibility*, London: Cassell.

Water Services Association (1997), *Waterfacts 1997*, Report, London: Water Services Association of England and Wales.

Wittfogel, K. (1957), *Oriental Despotism*, New Haven: Yale University Press.

3

The Role of Water in an Unequal Social Order in India

Deepa Joshi and *Ben Fawcett*

Introduction

Analysing social relations in contemporary India, Dube (1996) identified that the caste system prevails, its boundaries and hierarchies articulated by gender. The structure on which this social system is based was defined in Vedic philosophical texts about four thousand years ago[1] and re-interpreted and made more rigid over the following two millennia. The interplay of water, caste and gender in Hindu society, in which water both is polluted by the touch of the impure and purifies those who are so polluted, is evident in the lives of women and men living in a group of Himalayan villages in India. This chapter describes the historical evolution of the beliefs that underpin today's Hindu society, and reviews the particular constraints faced by Dalits and women.

Religion and the Basis of the Caste System

In the socially graded caste system, Brahmans are considered the purest, due to their traditionally prescribed involvement in scholastic, ritual and religious activities, tasks that are considered socially superior. The Sudras, at the other end of the social spectrum, are identified as permanently defiled as a result of their occupational engagement in polluting activities (Murray 1994). Several authors claim that although social hierarchy was established in the early Vedic period, social organization was not rigidly defined by caste; caste-based occupations were interchangeable and did not restrict social intercourse.

A rigidly closed caste-system, based on inheritance and determined by ritual notions of purity and pollution, evolved later when the core concern of the Hindu social structure became the maintenance of caste purity (Dumont and Pocock 1959). There are essentially two ways to bring about a condition of purity, one is to distance oneself from objects that signify impurity and the other is to purify oneself with things recognized to have the ability to absorb and remove pollution directly.

Temporary impurity is ascribed to the organic aspects of life, symbolized by the peripheral extremities of the human body, including the physical margins and matter issuing from them, (including) hair, nails, spittle, blood, semen, urine, faeces and even tears (Das 1982, Murray 1994). Human bodies in the act and process of producing bodily secretions or associating with these materials are polluting. Thus all women, regardless of their social caste, become polluting during menstruation and childbirth. Men also acquire this impurity during birth and death; however, while birth signifies auspicious impurity, death is associated with inauspicious impurity (Das 1982). In contrast, permanent impurity is ascribed to the Sudras due to the imposition of all-defiling, polluting tasks on this particular group. Sudras have historically been assigned the tasks of cremating the human dead, handling dead animals, handling human faeces and cutting hair and nails as well as the washing and cleaning processes associated with bodily excrements and are eternally impure (Dube 1996).

Water is considered to have intrinsic purity and the capacity to absorb pollution and to carry it away (Murray 1994). Water is also very prone to pollution, by touch or association with things and people considered impure. However, purification applies only to temporary impurity; Sudras cannot be purified. Equally importantly, impurity as a result of caste lineage is irreversible. Chakravarti notes that the mixing of castes (*varnasamakara*) violates the fundamental principle of Hindu social organization, because 'caste blood is identified as bilateral, i.e. received from both parents'(1993: 273). Until very recently, the severest condemnation and greatest legal punishment were applied for inter-caste contamination, especially that involving a Sudra with any of the upper-castes (Chakravarti 1998). Though denounced legally, the institution of caste continues to pose serious problems in the restructuring of Indian society, as traditional discriminatory practices persist into the twenty-first century (Jaiswal 1998).

The Brahman social order, based on the implied morality of women, is achieved in the patriarchal institution by the social, economic and political coercion of women. This is conveniently achieved through women's consent, as such unequal practices are internalized as Stridharma, the moral obligations on women cited in religious texts, required to secure good rebirth and/or eternal salvation.

To analyse the role of water in the regulation of a hierarchy based on both caste and gender it is necessary:

- To trace the origin and evolution of the belief systems involved;
- To review the current social and cultural contexts to assess whether the institution of caste still exists and, if it does, how it defines gender relations and water use practices.

The Vedic Period

The Sanctification of Water In early Vedic texts, water is referred to as apah and considered to be purifying. 'Hail to you, divine, unfathomable, all purifying Waters' (Rg Veda). In the use of water in daily life, as well as in ritualistic ceremony, the Vedas identify water as the very essence of spirituality, 'the first door to attain the divine order' (Atharva Veda). For example, during the act of taking a bath, it is not the water itself, but coming into contact with the sacredness of water which enables the attainment of knowledge that makes one sinless and guides one to the Eternal Self. 'Whatever sin is found in me, whatever wrong I may have done, if I have lied or falsely sworn, Waters remove it far from me' (Rg Veda). Water is considered sacred, but one does not pray to water, the physical entity, but to the source of life and spirituality within water. Water is considered both purified as well as the purifier, and the real and spiritually conceived source of life (Baartmans 1990). The concept of purification is thus essentially spiritual, rather than moral and/or physical.

The Institution of a Social Order The principles of social stratification in human society, on the basis of colour, class, individual capacity, occupational aptitude and moral and intellectual worth, were established in the early Vedic period. 'Rg Vedic Society witnessed a continuing struggle between the fair skinned *Aryas* and the indigenous tribes [referred to as slaves], who were viewed with particular hostility for their dark skins and racial inferiority' (Chakravarti 1993: 277). The social division of mankind (sic) into four *varnas* from the Purusa or the Eternal Man is described in the hymn *Purusa Sukta* of the Rg Veda: 'When they divided Man, how many did they make Him. The Brahman was his mouth; his arms the Rajanya (Kshatriya); his thighs were the Vaisya; from his feet the Sudra was born.' Social hierarchy was seen as divinely ordained. The Rg Veda also defines designated occupations in relation to *varnas*, 'One to high sway (Brahmana), one to exalted glory (Kshatriya), one to pursue his gain (Vaisya) and one to his labour (Sudra).' Brahmans were to be the teachers of mankind, Kshatriyas were to carry weapons and protect people, Vaisyas were to provide food for the people and the Sudras were to be the footmen or servants of the other *varnas*, even if they had all originated from the same Eternal Man (Prabhu 1939).

Despite the obvious social stratification, historians think a rigid caste system did not exist in the early Vedic period; the *varna* system is identified rather as an open class system of flexible membership. The concepts of untouchability, physical purity and pollution, and the prevention of social relations between individuals of different *varnas* are not evident in early Vedic literature (Kane 1974, Assayag 1995, Jaiswal 1998). It was later that inheritance-based rights and privileges came into practice (Kane 1974, Crawford 1982).

Gender Relations A popular school of thought argues that the evident gender bias in Hindu cultures today does not reflect the 'high respect accorded to women' in early Vedic civilization. For example, Polisi quotes early Vedic texts referring to 'equal roles for women and men in work, both religious and secular; ownership (capital) rights of women and women's ability to plead their own court cases' (2003: 1). Similarly, she cites the higher power and strength of Hindu goddesses in comparison with male gods, as evidence of greater female power or Shakti. Gender inequality is said to be an outcome of subsequent male interpretations of traditional texts and, later, from the impact of Islamic subjugation and the colonial capture of India (Polisi 2003: 1). Meanwhile Chakravarti (1993) points out from Lerner's (1986) archaeological studies that the most egalitarian societies were the hunter-gathering tribes in Central India in the Mesolithic period (c. 12,000 to 8,000 years ago) amongst whom, women and men's roles were separate but equal and interdependent. Women's reproductive role was thus valued as much as, if not more than, their productive contribution.

Practices of caste, gender and class stratification became established in India during the evolution from a hunter-gathering to a pastoral and then eventually agrarian society. The Rg Vedic society was agrarian and displayed a culture of organized labour in production, through the capture and subjugation of the 'indigenous people'. This enabled the labour of Aryan women to be restricted to the household.

Chakravarti quotes from the early Vedic text Satapatha Brahamana to demonstrate the evident control over women's sexuality: '... the fears that the wife might go to other men; ... the divine raja, Varuna, seizes the woman who has adulterous sex with men other than her husband' (1993: 278). Chastity was thus identified as a feminine virtue. Evidence of a patriarchal society was also indicated by the spatial and social adoption of the husband's home and family by the wife. These examples illustrate the process of evolving patriarchy, at the cost of female subjugation and inequality. Finally, the few eulogized examples of women scholars and spokespersons in this period were an exception rather than the rule.

The Post-Vedic Period

There is little dispute amongst researchers concerning the rigid social stratification seen in the post-Vedic periods (500 BC to AD 300), which was achieved through both a caste segregation in food and water use habits, and women's subordination in the domestic environment.

The Smrtis, a body of post-Vedic literature, describes the role assigned to water in determining ritualism in relation to bodily purification. Ritualistic purification practices involving water are firmly recognized as contributing to Dharma or moral law and the most authoritative text on Dharma is the 'Laws of Manu', or *Manusmrti* (Crawford 1982). Manu is often blamed for legalizing social stratification by caste and gender; however, some authors argue that he may have simply recorded the system of social order that existed (Kane 1974). Nevertheless, Manu and others recorded and so codified the social order as a set of morally appropriate behaviours, duties and/or obligations. The religiously inclined Hindus tenaciously practise Dharma in the hope of good rebirth and/or salvation despite the fact that there is no watchdog, as in the Western Church, to enforce moral regulations. The notion of Dharma, signifying moral conduct, persists in Hindu society (Nagarajan 1994).

The Established Principles of Caste and Gender Murray (1994) lists some of the structural features of the caste system presented by Manu:

- A rigid caste status assigned solely on the basis of inheritance;
- The Brahmans' uncontested prominence in cultural and religious ritual;
- The Sudras' eternally polluting status;
- The centrality of a person's caste in his/her social life; prohibition of mobility across caste boundaries, maintained principally by the regulation of intra-caste marriage and, in practical terms, through water-use and eating arrangements;
- Extensive norms and elaborate rituals prescribed for regulating social stratification;
- Enormous social energy devoted to maintaining caste boundaries based on the concept of Dharma.

The post-Vedic law books or Dharmashastras define clearly how Sudras are essentially and irreversibly allocated polluting tasks. In order to maintain purity, the Sudras were excluded physically, socially and morally from the larger village commune. Fa Hein, a Chinese traveller in India, writes about how, in a public place, the Candalas (those tasked with ritual cremation of the dead) had to give notice of their approach by striking a piece of wood to warn others to avoid contact with them (Kane 1974). Purification if touched by a Sudra involves taking a cleansing bath; after talking to a Sudra one is purified by talking to a Brahman; and after seeing a Sudra one is purified by looking at the sun, moon or stars and rinsing one's mouth with water (*acamana*). Water is the primary medium for removal of pollution obtained from the Sudras.

It is explicitly stated that a well, or any other source of still water, is polluted by a Sudra's touch. Manu elaborates the rituals to be performed to purify such polluted water (Khera 1997). Water, and food cooked in water, offered or touched by the Sudras are polluting. According to the Dharmashastras, the Sudras are excluded from all religious knowledge and ritualism – the very basis of Hindu existence. The Dharmashastras provide numerous parallels between socially belittled animals, like dogs and pigs, and the Sudras who are literally isolated from the other caste groups, and contextually removed from the class of humans. According to Manu, all persons become polluted and therefore 'untouchable' during birth and death in the house. People who touch those mourning during a death, touch the corpse or those carrying the corpse to the cemetery are also polluted. Drawing parallels with pollution in death, he likens Sudras to 'a living cemetery'.

Women are cyclically impure, during menstruation and after childbirth, regardless of caste. The social blight accorded to (especially upper-caste) women during menstruation is legitimized philosophically by comparing the menstrual cycle to the three-phase cycle of Hinduism: creation by Brahma, maintenance or preservation by Visnu and destruction by Siva. During the first four days of menstruation – the damaging *tamas* phase – women are believed to have powers capable of reducing the life-span of men and vegetation; hence the need for seclusion and the danger of pollution (Kondos et al. 1990). The next twelve days are the *rajas*, or reproductive phase, and the remainder of the cycle is *sattva*, which is neither creative nor destructive. This three-phase structure also applies to the cycle of pregnancy. All of the restrictions detailed for the Sudras apply to all women during menstruation and after childbirth.

Upper-caste men who touch Sudras, a corpse, menstruating women, or women during the first ten days after childbirth, are considered polluted, the remedy to purification being bathing (Das 1982). However, men are considered relatively pure, as opposed to women, who, given their regular and periodic pollution, are identified as fundamentally impure (Hershman 1974 quoted by Krygier 1990). Water is the essential medium to purify all forms of temporary pollution.

Several post-Vedic texts indicate that marriage signifies the beginning of *real* existence for the Hindu woman. A woman takes up permanent residence at the husband's home and secures for the first time her own home – prior to that she is only a temporary guest in her father's home. Several strictures are levied on fathers to give away their daughters (considered the highest sacrifice with maximum spiritual gains) by the age of eight so that women can be the sexual property of their husbands at the dangerous time of puberty and there is less possibility of women going astray (Chakravarti 1998). The moral code of conduct for women, Stridharma, prescribes women's virtue in domesticity, as well as their control of their own sexuality. Any physical association of upper-caste women with Sudra males is subject to severe punishment, leading to castration for the Sudra (even if the association was mutually desired) and/or death for both parties.

Upper-caste women (Brahman and Kshatriya) obtain their caste citizenship and its moral and spiritual benefits only on marriage.[2] The female world initiated through marriage is limited to the household, which is the focal point of female reproduction, domestic labour and patrilocal kinship relations (Chakravarti 1998). *Stridharmapaddhati*, an eighteenth-century Maratha text which gave a

legal reformulation of Manu's writing, prescribes the code for the 'perfect wife':
'Obedient service to one's husband is the primary duty enjoined by the sacred
tradition for women' (Chakravarti 1998: 273). Virtuous women combine the dual
action of daily household chores and assisting[3] the husband in his religious duties.
Schooling in *stridharma* kept women's assumed libidinous nature in check.

While there is a general belief that Hindu women are inherently more impure
than Hindu men, upper-caste women are identified as relatively less impure than
lower-caste women, who are prescribed a longer time for bodily purification.
This is primarily related to the practice of female domesticity and the high values
accorded to the use of water and food preparation. The kitchen was probably the
most sacred place in the higher-caste Hindu house (Dumont and Pocock 1959).
It was the place where a ritual break was made in purifying food contaminated
in production (by those of other castes) before it was admitted to the body. The
hands providing this break needed to be those of higher-caste women (Douglas
1970).

Numerous references are made in post-Vedic texts to the need to restrict upper-
caste women to the kitchen. By contrast, productive roles are assigned to lower-
caste women, exemplified by such working couples as the barber and the midwife,
or the washerman and washerwoman (Krygier et al. 1990). This does not, however,
imply that their husbands shared the domestic burdens with them.

Compliance of women with their domestic role and subdued sexuality was
considered as wisdom; service to one's husband equated with worship of god. Compliance to the moral code held rewards: virtuous women gained a good reputation
and immediate happiness in this world and went to heaven after death.

Social Stratification in Contemporary Hindu Society

Questioning whether historical interpretations of local culture are valid, Sax
(1991) compares the dualities of Hindu history, of sacred and profane, of body
and spirit, to Trojan horses, determined historically and only culturally specific.
Srinivas (1998) supports this argument by pointing out that economic, political
and Western axes of power determine dominance. The agriculturally landed and
the numerically strong are the currently dominant social groups in rural India and
this does not include the Brahmans. Others argue that despite reform, legislation
and wider potential for choosing occupations, socially allocated roles prevail in
that no one else but a Brahman can perform the function of the priest, while the
Sudra remains responsible for ritually-polluting occupations (Das 1982, Dube
1996). This implies that while social positions may have changed for the more
and less dominant social groups, little has changed in the social order of Dharma
in terms of both culture and ritual. Brahmanical patriarchal practices, established
through both caste and gender hierarchies, and rendered invisible by being
defined as religion, custom, tradition and honour, continue to contribute to social

stability and salvation. Assayag (1995) reports recent adaptations where higher-caste Hindus confessed to removing the strictures of caste in the outside world and donning these again at home.

Rituals and Social Exclusion Through Water in Chuni Village[4]

Chuni is a breathtakingly beautiful village in the Kumaon hills of the Pithoragarh district, in the central Himalayan state of Uttaranchal. The village is located at the top of a hill, a one-and-a-half-kilometre steep ascent from the valley of the Ram Ganga river. Research in the village yielded a valuable understanding of current beliefs and practices in relation to caste, gender and water in this rural Hindu context and indicated how little has changed from past centuries. The account below analyses:

- Unequal social relations in Chuni and nearby villages;
- Gendered roles and responsibilities in relation to the use and management of water;
- The social order of water management.

Social Relations in Chuni

There are approximately fifty households in Chuni, belonging to a clear hierarchy of five caste groups and seven social groups within those. The Kshatriya families head the caste hierarchy in Chuni, including two Khanka landlord families who are dominant, and two other families settled by the Khankas. A single Brahman family group, the Joshis, although of higher caste than the Kshatriyas, is of lower class, having been settled by the dominant Khanka family to till their land. They strive to maintain good relations with the Kshatriya families. Two mixed-caste family groups are of lower-caste and low class, and are unable to exercise power in the village. They were also brought in by the Khankas to till the land. The Sudra or Dalit[5] families, at the bottom of both the caste and class hierarchies, were traditionally blacksmiths, artisans and musicians and occasionally provided agricultural labour. The Dalits, as others, were settled by the dominant Khanka family. As elsewhere in Uttaranchal state in Chuni, the Dalits are a minority and this reflects their occupational need to spread out.

Characteristics of post-Vedic ideology are active in Chuni. Appropriate social behaviour and social relations are mediated through caste-based norms. Caste units are specifically endogamous and social ostracism is severe, especially if one of the partners in an inter-caste union is a Dalit.

Hira Devi Tamta, a Dalit woman, married a Brahman man when she visited her sister in Punjab. She and her husband are not accepted in his ancestral village. Hira Devi's husband lives and works in Punjab and only visits his family in Chuni from time to time. The dominant Kshatriya caste in Chuni denounce him as a social outcast, but blame Hira Devi for her loose morality in having lured a Brahman.

In contrast, a Brahman mother of four young children eloped with a young unmarried Kshatriya man in nearby Himtal village. The couple was brought back home and the Brahman family were temporarily ostracized. The Brahmans, who were dominant in the village, argued that the Kshatriya boy had performed black magic to lure the woman and did not readily forgive him. It pays to believe that Brahman women are usually not of loose morality.

The relative severity and laxity of social ostracism in these two cases, show that power structures within the community determine who is to blame and how much they should suffer for their behaviour.

Elderly Kshatriya women in Chuni are very conscious of the practice of commensality, in which water and food are only accepted from those castes that are of the same social status or 'higher'. These women often invited the researcher, by caste a Brahman, to their kitchens for meals, saying that they considered it a moral virtue to feed a Brahman, albeit one who was considerably younger than them. Srinivas (1994) reports that the practice of caste taking precedence over age or class, in determining social respectability and status, is common.

After a few days, news of the researcher's eating and smoking with the Dalits spread across the village. This news disturbed the hospitality of the Kshatriya women but nothing was asked and nothing explained. One day, Hima Devi, a Dalit woman harvesting the fields belonging to a Kshatriya widow, Saraswati Devi, offered the researcher a *roti* (flat bread) in Saraswati Devi's presence. The offer of food was essentially to prove to Saraswati Devi that not all Brahmans practise exclusion. Hima Devi gained her small victory when the researcher accepted and ate the bread in the presence of Saraswati Devi.

Subsequently, the researcher's relations with Saraswati Devi remained cordial, but Saraswati Devi refrained from calling the researcher again to her kitchen and also avoided all physical contact.

In Chuni, the Dalits live in hamlets isolated from the main village, where the 'upper-castes' live and share a common *naula* (an underground source of spring water, the traditional form of water supply in the hills). The Dalits have a separate *naula*.

The Dalits cannot enter the inner confines of temple spaces, nor do they have equal access to local rituals. In the mountains, where religion is a central part of people's lives, this has resulted in the Dalits establishing separate local gods and separate religious rituals. However, the keepers of the dominant religious beliefs, the Brahmans, consider all such practices as incomplete and a mockery of 'true' practice.

The Effect of Counter-caste Legislation

The effect of generations of living in social obscurity is evident in the physical postures taken when two people communicate. Tully (1991), writing from his extensive experience in India, describes a conversation between a Dalit and a high-caste Hindu during which the Dalit folded his hands and wagged his bowed head from side to side. Examples of behavioural differences are evident in the hill villages today, although, to some extent, moderated by legislation.

A group of 'upper-caste' men and women in nearby Pilkhi village remark: 'There was a time when these scheduled castes bowed down low, their foreheads nearly touching the ground. Today they barely greet us. It is we who are the oppressed today, by this whole context of legislation against caste.'

However, social control is still evident. A good Dalit is one who bends, keeps a distance, keeps away from main paths and does not touch water sources. If Dalit men or women speak their mind and try to overcome their inequality, they are labelled as having challenged the socially acceptable limits.

An upper-caste woman, just married and young, new to the village and therefore of low social stature still exudes a natural physical defiance in front of a Dalit man of the same village. If the Dalit man is a guest in her house, she will not wash his used cup or plate even though washing utensils is her task in honour of the visiting guests, especially when they are males.

A young local lawyer said, 'The number of cases registered under the Harijan Act does not give evidence of all the incidences of oppression or social revolt. Far too many cases do not reach the courtrooms. They are settled in the village either through negotiation or threat. This is characteristically true in the hills, where the Dalits live as a minority' (interview Joshi 2000). Nonetheless, constitutional amendments have, to some extent, reduced caste-based inequities: a Dalit woman, asked what changes she had experienced in *her* life as a result of counter-caste legislation, pointed to her husband and spoke of his new-found manhood which he had not known only ten years earlier (Tully 1991). This highlights the fact that gender disparities within caste hierarchies remain largely unidentified and have not been addressed.

The Social Control of Women

Sanctification of women as mothers, and graceful and honourable wives and daughters, is evident in Chuni. Accordingly, women are both compelled and enticed to maintain the dignity of the family, local community and society. The social pressure to maintain this dignity is as powerful as the imposed self-determination on both 'good' and 'bad' Dalits. Two illustrations of the links between gender and caste are outlined below:

- the differential access to and control of agricultural land, which is the primary local resource;
- the assignment of increasingly difficult roles and responsibilities in relation to domestic water, as feminine.

Control of Land Agriculture is the mainstay of rural livelihoods and while women form the backbone of the agrarian economy in the mountains, as in much of India, they do not own land. Consequently a woman's status is reduced to that of a dependant and, according to her marital status, she either depends on the benevolence of her parents or on her husband and his family.

The notion of appropriate masculine and feminine roles and responsibilities is strongly established. However, men are often no longer the agricultural bread-winners. Outmigration of able-bodied men is a common survival strategy in the mountains, resulting in women being pushed from their homes into the fields. Given the social history, this increased mobility is assumed by many observers to indicate the 'better social status of the hill women'. Yet, while most masculine jobs, with the exception of ploughing, considered an inherently masculine task and highly inauspicious for women, are readily handed over to women, agricultural practice remains predominantly patriarchal with the ownership and decision-making remaining with the elderly patriarchs or absentee males (Mehta 1993).

Radha Devi is a young Brahman woman of around thirty years of age. She was physically tortured by her in-laws for the fifteen years of her married life. Finally forced out of the house, she sought refuge in her father's home. Here, she lives with her youngest son in a small hut, which was earlier the family kitchen. Her widowed mother offered her some land to till and a cow, but she has not been given formal ownership rights over these possessions. According to her mother and brothers, this is a temporary arrangement and she has to return to her husband's house when her son is able to claim his right to his father's land. Her father's land belongs to her two brothers, both of whom have migrated away from the village. Radha Devi lives here as a stranger, constrained by the obligations made upon her by her own family.

In the four villages in which this research was done, in only two instances had land been given to daughters for their use and they did not have authority or control over that land.

In Chuni the Khanka patriarch gave away some of his land to his daughter, but registered it in the name of her husband when his sons' wives did not bear any male children. The Khankas regretted this decision when their daughters-in-law bore sons later; to this day, this household traces its present-day poverty to the hasty act of giving land to the daughter's family.

Women's Water Roles In their socially determined roles as home managers, and without the option of migration, women carry the brunt of the struggle for survival. Throughout the year most women work for fifteen to sixteen hours each day, collecting water, fodder and firewood, cultivating the fields, depositing animal dung everyday in the fields, cooking, washing, cleaning the house, and looking after the cattle, the children, the old and the sick. The Hindu belief in self-regulation means that good women are those who perform all these designated tasks willingly and diligently, ensuring the wealth and prosperity of the household.

Married women carry much of the responsibility for providing and managing water for the household. This gendered role does not diminish with age for women, while men's responsibilities – but not authority – often lapse in old age.

One of the first rituals performed by a bride, when she enters her husband's home, is to fetch water for the family from the local *naula*. Culture demands that the groom should accompany the bride, but in the majority of cases, the groom's paper crown, used during the wedding, is tied to the bride's head, instead, signifying that his visit is only symbolic. The burden of water-related work, commonly assigned to young women during adolescence, is thus cemented by marriage. From then on, water-related tasks are a responsibility, not a choice. As Anandi Devi of Chuni says: 'There is nothing better than *jawani* [youth]; to throw a scarf around your neck and skip along crooked paths. Yes, there is work, but it is not binding. My daughter, Janaki, may cut some grass and fetch some water today, but she certainly doesn't have to worry how to feed the family day and night.'

Both boys and girls help meet their families' water needs, but the responsibility for fetching, using and managing water is increasingly imposed on girls as they grow up and boys learn as adolescents that fetching water and water-related tasks are not their responsibility.

Relief from assigned water responsibilities for adult women happens only during menstruation, when they are considered physically impure. Kshatriya women say: 'We become Domnis [Dalit women in the condescending local parlance] then, barred from water sources and homes.' The Dalit women become doubly impure.

Factors which contribute to the social exclusion of 'impure' women vary in intensity between villages. In most mountain villages, women when impure live in small huts outside the domains of the main house. 'Impure' women are excluded from the *naulas* in all villages. During such periods, women do not fetch drinking water or cook food, and remain dependent on other members of their family to perform this work. However, they still perform tasks like washing clothes, and fetching water for uses at home that are not considered polluting; for these they have to get water from other sources, such as storm water drains and rivers. Drains are not considered sacred, and rivers, though sacred, are believed to absolve all pollution because they flow continuously. If such alternative sources are not available, women depend on others for all their water needs.

Despite their central responsibility for fetching and using water, women, regardless of their caste and class, have historically remained excluded from decision-making in the planning and management of water supplies through the tradition of female seclusion. Such decision-making is public and socially identified as masculine, the domain of powerful men.

However, locally, neither women nor men equate the gendered division of work with inequality. In practical terms, sharing water responsibilities with men or *interfering* in water management decisions would lead to ridicule, as the carefully preserved image of good women would be eroded and the dent in (certain) men's masculinity would be blamed on women. Women argue that, no matter how difficult their tasks in accessing water, their water roles nevertheless grant them some degree of social value in what they identify as a demeaning existence.

The Constructs of Social Hegemony   Although some men assist their wives and mothers at home, when the local constructs of male hegemony are violated, the man, and even more so his wife, become the object of ridicule.

Khim Singh and Kishore Singh are widely recognized as men who have gone beyond what is appropriate. Disapproval relates especially to their fetching water, cutting grass, milking the cattle, and collecting animal dung for the fields; all of these tasks are defined as being exclusively feminine. Women themselves blame the wives of these men: 'What a shame to sit and watch their husbands work. Their homes are *destroyed* and so will be their daughters, who learn from them.' Observation reveals that the few men who assist their wives' domestic work are usually those in paid employment; perhaps because they have already fulfilled the masculine notion of being the breadwinners and therefore have no fear of being belittled.

Anadi, a young, male Dalit teenager, assists his blind mother in fetching water. Anadi's mother is not proud, but ashamed of this. 'It is my greatest misfortune that I bring this upon Anadi. His sisters are both married and everyone looks down on him in our poverty and on him helping me with the housework.' A gender role-reversal caused by poverty and sharpened by caste makes Anadi a typical case of 'a lesser man'. Normally acquiring a paid job or marriage are important landmarks that define the start of masculine roles and responsibilities.

In the local context, gender role reversal, far from signifying equality, indicates socio-economic impoverishment and hopelessness. It is the poor amongst men, like Khim Singh in Chuni, who sometimes carry water for richer households. In contrast with women, men are paid in cash and kind for such work; yet it is still considered demeaning and Khim Singh would never carry water for his own home, so great is the pressure on him to restore what is left of his 'masculinity'.

Poverty here is seen as a harsh leveller of gendered inequities, as self-esteem is low amongst the very poor women and men who have reversed traditional roles. A woman who works as a daily-wage labourer in road construction said, 'What difference is there between a dog's life and mine? I work here, abused and ridiculed by the male contractor and his managers and then, when I go home, I have to do everything myself.'

Water, Caste and Gender – The Matrix of Inequality in Chuni

Chuni is a water-abundant village. The temple to the Jal Devi (Water Goddess) in the village is said to keep the waters flowing in the small agricultural drains (*guls*). The traditional water springs, or *naulas*, are said to be the abode of the Jal Devi and are worshipped. Eulogizing the appropriateness of traditional systems, Agarwal and Narain (1997) write of the *naulas* being held in deep reverence traditionally and of the rituals performed while constructing these systems.

> Dalit men have traditionally built the stone walls that protect a naula, given their artisan skills. However, the ritual of pollution, ascribed to Dalits, implies that the structure in sources used by the upper-castes has to be religiously purified (with water!) after the construction is complete. Thereafter, Dalits and also women of all castes, when cyclically impure are not allowed access to these sites.

The self-regulatory mechanism works through propagating the belief that serious harm, including mental derangement, befalls those who violate these norms. Women in their cyclically impure state are often blamed for polluting the water sources, reputedly resulting in depletion of the flow of water in the *naulas*. Purification is then required and both women and men say: 'Big white snakes appear in the *naulas* as the first sign of the wrath of the Goddess, and then the water slowly dries up. The remedy is a purifying Devi Path [chant] and also presentation of a calf to the Brahman performing the ritual.'

Officially, the Kumaon and Garhwal Water Act of 1975, terminated the customary ownership and usage rights of individuals and village communes and, according to the Act, the 'State took over the power ... for collection, conservation and distribution of water and control of water sources'. However, in the village, specific hamlets and/or families exercise control over these formally 'state-owned' *naulas*, as well as other water sources, and they determine who can access them (Rangan 1997). Local culture excludes the Dalits in Chuni from using any naula in the village except the one that is assigned as theirs. Ganga Devi, a Dalit woman, describes how her small nephew was recently beaten for *stealing* cool water on a hot summer day from the *naula* belonging to a mixed-caste group. To purify the *naula*, defiled by the touch of this small Dalit boy, the collected water was thrown away. The 'owners' performed a ritualistic prayer and warned the boy's parents that if the act was repeated they would no longer be given land for sharecropping.

In a nearby village, a Dalit schoolteacher, not from the local area, was socially ostracized and finally forced to leave the village for his defiance in fetching drinking water from the *naula* used by the upper-castes. This was not tolerated, especially as there was a tap which villagers recognized he could safely use. Taps provided through official projects, in contrast to *naulas*, are not sacred. However, water from taps mostly comes from storm water drains, and is of inferior quality and taste. Locally, this defiance was seen as an example of what happens when the Dalits become educated: 'It makes them perverse.' Despite legislation that makes such treatment a criminal offence, even educated Dalits like this schoolteacher do not pursue legal action. In the mountain villages, Dalits, living as a minority, are well aware of what suits their specific interests better.

Hira Devi, of Chuni, said, 'What can we do? Will you come with me to the *naula* and I will openly take water from the Khanka's *naula*?' However, after much thought and discussion, she dropped the idea. She said, 'My neighbours are important to me, no matter what they do. I need their support for my family's daily existence. You sympathize with me, but you are here today and gone tomorrow.' Conflict and defiance even in the face of social injustice was not her preferred solution.

The degrees of social hierarchy in Chuni closely parallel the relative access to preferred sources of water and the adequacy of those sources. Dalits and the other service providers are settled in the least attractive peripheries of the village.

This explains why some upper-caste women in some parts of Chuni could say: 'We are water lords [sic] here'; while Dalit women in the same village said: 'Ask us what water scarcity is? It is not to bathe in the summer heat after toiling in the fields. It is to re-use water used in cooking for washing utensils, to use this water again for washing clothes and finally to feed the soapy water to buffaloes. It is to sit up the whole night filling glass by glass as water trickles from our *naula*. It is to wait for someone from the Khanka household to give us water from their *naula*. It is to walk up and down their path, calling a little, waiting a little, hearing them say they are too busy and helplessly remembering our own tasks at home. It is to *steal* water stealthily, taking care not to spill any on the concrete floor for fear of being suspected; feeling the guilt of stealing. It is all this and much more; being obliged physically, socially and morally for the water they give us from their *naula*.'

The eight Dalit families in this village have access to one *naula*, while one Khanka family has exclusive, legitimate access to one *naula*. Even when numerically equal, inequality persists on the basis of untouchability. In remote mountain villages, traditional sources like the *naula* are still widely used. Water supply systems provided by government departments either have not reached them or, where provided, are unreliable, poorly maintained, and/or considered inappropriate for the multiple household water needs. In the Kumaon hills, Dalits remain excluded from contact with the authorities that make decisions about water in the village. Similarly women, barred from the public domain by social norms, remain excluded from contact with decision-makers so the design of water systems has to date not met the specific needs of Dalit women in particular.

Recent policies suggest that giving authority to local communities increases the sustainability of water delivery systems and improves access. The management of projects for the 'improvement' of water supplies is assigned often to an assumed homogeneous community. The impacts of such policies, which lack an informed understanding of the local social systems, are discussed in Chapter 8.

Conclusion

The Hindu philosophical belief that both water and the human body are social constructs persists. In the rural mountain villages, water is instrumental in determining inequality. Consequently, the Dalits remain permanently excluded and women are cyclically excluded from traditional, but preferred, water supplies. This case study suggests that legislation has resulted in few positive outcomes with regard to reducing the severe constraints on women, particularly lower-caste women. While the details may vary between villages, the universal beliefs about ritual purity and pollution and the role of water in defining these suggest that the situation may be similar wherever the Dalits are a minority.

Murray (1994) wrote that the status of some could be raised only by lowering the status of others; to decrease the impurity of untouchables and the control on women's sexuality has the long-term consequence of challenging and eroding notions that are central to the Hindu way of life. Skirting around these issues will at best result in cosmetic changes to the social fabric of Hindu society and the related distribution of, and access to, basic resources.

Notes

1. Vedic philosophy refers to ethical thought presented in Vedic literature – the Vedas and the later Brahmanas – written approximately 2500–600 BC. The precise period is disputed.
2. This principle is applicable to the *dvi-ja* 'twice-born' community (Brahmans and Kshatriyas). Men reach this spiritual state through the critical *upanayana* (thread-wearing) ritual, which takes place in early adolescence.

3. All forms of domestic ritual in Hindu worship are ideally performed by a man and his wife. The man grants a woman the privilege of participation, but women always play a subordinate part.
4. All locations and individuals used in the case study have been given new names to preserve anonymity.
5. Officially, the Sudras are known as the Scheduled Castes. Mahatma Gandhi termed them 'Harijans' or 'people of God'. Politically, this terminology is strongly rejected by the Scheduled Castes, who call themselves Dalits, or 'the oppressed'.

References

Agarwal, A. and S. Narain (1997), 'Dying Wisdom: Rise and Fall and Potential of India's Traditional Drinking Water Harvesting Systems', *State of India's Environment – 4, A Citizen's Report*, New Delhi: Centre for Science and Environment.

Assayag, J. (1995), The Making of Democratic Inequality – Caste, Class, Lobbies and Politics in Contemporary India (1880–1995), *Pondy Papers in Social Sciences*, 18, Pondicherry: French Institute of Pondicherry.

Baartmans, F. (1990), *Apah, the Sacred Waters: An Analysis of a Primordial Symbol in Hindu Myths*, Delhi: BR Publishing.

Chakravarti, U. (1993), 'Conceptualising Brahmanical Patriarchy in Early India: Gender, Caste, Class and State', *Economic and Political Weekly*, 3 April, Mumbai: Sameeksha Trust.

—— (1998), 'Law, Colonial State and Gender: Men, Women and the Embattled Family', *Rewriting History: The Life and Times of Pandita Ramabai*, New Delhi: Kali for Women in association with The Book Review Literary Trust.

Crawford, S.C. (1982), 'The Evolution of Hindu Ethical Ideals', *Asian Studies at Hawaii*, 28, Hawaii: University of Hawaii.

Das, V. (1982), *Structure and Cognition, Aspects of Hindu Caste and Ritual*, New Delhi: Oxford India Paperbacks.

Douglas, M. (1970), *Purity and Danger*, London: Penguin Books.

Dube, L. (1996), 'Caste and Women', M.N. Srinivas, ed., *Caste: Its Twentieth Century Avatar*, London: Penguin Books.

Dumont, L. and D.F. Pocock (1959), 'Pure and Impure', *Contributions to Indian Sociology*, 3, London: Sage Publications.

Jaiswal, S. (1998), *Caste: Origin, Function and Dimensions of Change*, New Delhi: Manohar Publishers and Distributors.

Kane, P.V. (1974), *History of Dharmasastras*, II(1), Poona: Bhandarkar Oriental Institute.

Khera, K.L. (1997), *Index to History of Dharmasastra by Pandurang Vaman Kane: Comprehensive Guide to Hindu Rites and Rituals*, New Delhi: Mushiram Manoharlal.

Kondos, V. (1990), 'The Triple Goddess and the Processual Approach to the World: The Parbatya Case, Caste and Female Pollution', in J. Krygier, M. Allen and S.N. Mukherjee (eds), *Caste and Female Pollution, Women in India and Nepal*, South Asian Publications Series (7): Asian Studies Association of Australia, New Delhi: Sterling Publishers.

Krygier, J. (1990), Caste and Female Pollution, in J. Krygier, M. Allen and S.N. Mukherjee (eds), *Caste and Female Pollution, Women in India and Nepal*, South Asian Publications Series (7): Asian Studies Association of Australia, New Delhi: Sterling Publishers.

Krygier, J., M. Allen and S.N. Mukherjee (eds). (1990), *Caste and Female Pollution, Women in India and Nepal*, South Asian Publications Series (7): Asian Studies Association of Australia, New Delhi: Sterling Publishers.

Lerner, G. (1986), *The Creation of Patriarchy*, Oxford: Oxford University Press.

Mehta, M. (1993), 'Cash Crops and the Changing Contexts of Women's Work and Status: A Case Study from Tehri Garhwal', *Mountain Population and Employment*, Discussion Paper Series (2), January, Kathmandu: International Centre for Integrated Mountain Development.

Murray, M. Jr (1994), *Status and Sacredness: A General Theory of Status Relations and an Analysis of Indian Culture*, Oxford: Oxford University Press.

Nagarajan, V. (1994), *Origins of the Hindu Social System*, Nagpur: Dattsons.

Polisi, C.E. (2003), 'Universal Rights and Cultural Relativism: Hinduism and Islam Deconstructed', *The Bologna Centre Journal of International Affairs*, Spring, Italy: The Paul H. Nitze School of Advanced International Studies.

Prabhu, P.H. (1939), *Hindu Social Institutions – With Reference to their Psychological Implications*.

Rangan, H. (1997), 'Property vs. Control: The State and Forest Management in the Indian Himalaya', *Development and Change*, 28 (1).

Sax, W.S. (1991), *Mountain Goddess, Gender and Politics in a Himalayan Pilgrimage*, Oxford: Oxford University Press.

Srinivas, M.N. (1994), *The Dominant Caste and Other Essays*, Oxford: Oxford University Press.

—— (1996), *Caste: Its Twentieth Century Avatar*, London, Penguin Books.

—— (1998), *Village, Caste, Gender and Method, Essays in Indian Social Anthropology*, New Delhi: Oxford India Paperbacks.

Tully, M. (1991), *No Full Stops in India*, London: Viking.

4

Naked Power: Women and the Social Production of Water in Anglophone Cameroon

Ben Page

Introduction

In 1959 a woman stripped off her dress and stood naked in front of a crowd during a political rally in the small Cameroonian town of Tombel. She was exasperated because, despite over twenty years of discussion about it, there was no piped water supply in town. Within a few years a new public network was in place and her protest had become part of the local story of constructing that water system. Three decades later the women of Tombel felt obliged to take to the streets once more as a protest against the government's attempt to close down public taps and charge users for water as it was collected. At the front of this recent protest a few old women marched naked. The next day the government's water engineers fled town, and they haven't returned since. This chapter is about power, and the capacity of women to develop water supplies despite their apparent powerlessness.

Throughout the literature on water and development the importance of women tends to rest on their role as water 'consumers' (van Wijk-Sijbesma 1998, Uitto and Biswas 2000). However, women should also be considered in discussions of the history of the 'production' of water. Women were not only the principal users of water, but they were also, in some ways, the makers of modern water supplies. Much of the existing literature ignores women's historic involvement in making water infrastructure, water institutions and water politics, rendering them invisible. Those who are now arguing that the projects of the past failed because they did not include women risk erasing a history of the ways in which some women achieved their objectives. Erasing this history of women's involvement is actively disempowering because it dismantles a tradition through which women have dragooned, coerced and persuaded men into participating in the production of water.

But what does the idea of the 'production of water' entail? Water and nature are so closely intertwined symbolically that it is counter-intuitive to talk of 'the production of water'. Water isn't made, it just exists. Despite the obviously social means by which humans acquire the water they need, the substance itself remains stubbornly natural to human eyes. Yet, the technological, industrial and economic process by which water is extracted from the ground or rivers, filtered and sterilized, distributed through a network, consumed, recollected, retreated and put back into rivers is a familiar one. Within that confined engineering context the idea of the production of water is unproblematic. But the idea of the production of water ought to include other elements too. It entails the social arrangements that govern the use of water, the rules that regulate the way that people behave around water sources, the committees that meet to decide about the allocation of resources, the local values that water is associated with and, finally, the cultural meanings associated with water (Strang 2004). All of these elements have a history and a politics; these combine to make water what it is in any particular place. The production of water stretches far beyond water engineering to encompass the broader relationship between water and society.

The production of water is a historical and geographical process that weaves together physical changes and changing ideas. Before water ever enters a pipe network and becomes an engineered product it is already a produced social substance. But equally as the water enters the pipe its social meaning changes; the material transformation of water and the social meanings of water are co-produced (Swyngedouw, Kaika and Castro 2002). This notion of the production of water draws freely on those authors who talk more generally about a 'production of nature' (Harvey 1974 and 1993, Smith 1984, Castree 1995, Swyngedouw 1997). They emphasize that cultural ideas about nature are often ideological and they reject any tidy separation of nature and society. To this is added the cultural theorization of commodity biographers, who argue that prior meanings adhere to objects long after the political and economic conditions of production have changed (Appadurai 1986, Kopytoff 1986).

The value of this expanded notion of the production of water is that it captures the myriad ways in which women are involved. By extending production to include elements such as the establishment of the rules of behaviour at a water source it is easier to see that women have always been a key part of this story. Even though women have been only minor participants in formal water engineering they have often made demands for changes in the physical infrastructure, access and the arrangements for payment. By using historic forms of protest (such as nudity) they have articulated their demands in an effective way despite having no formal control over decision-making. It is hard to make generalizations about gender relations in an area as culturally diverse as anglophone Cameroon, but in general most overt social structures suggest that men hold a dominant position. Yet there is still scope for women to find ways to express their own interests; there are established rituals of resistance. Key texts on gender in Cameroon emphasize the need for women to become more influential (Endeley 2001, Fonchingong 1999, Forje 1998), but they also show that despite their low status women can wield power (Ardener 1975, Diduk 1989 and 1997, Goheen 1996). This discussion draws out the ways in which these two aspects are linked in the field of water supply.

The empirical material was gathered from archival sources and interviews in Cameroon in 1998, 1999 and 2003 (Page 2000 and 2003). After a brief description of the physical context, the chapter looks at the history of women's involvement in producing water supplies during the twentieth century. This is divided into four sections: the pre-colonial period, the colonial period, the early post-colonial period and the present. Emphasis is placed on rural and small-town water supplies.

Water in Anglophone Cameroon

Anglophone Cameroon comprises the North-West and South-West, two of Cameroon's ten provinces. The South-West has very high rainfall (more than 3,000 mm/year) and the North-West has marginally less, but is higher in altitude. In general both provinces have high relief and are well watered, enabling most water supplies to be provided by gravity from springs. However, there are places (such as Tombel) where the lack of a drinking water supply limited pre-colonial settlement and later urban growth. Nevertheless, relative to many other regions of Cameroon, and indeed much of sub-Saharan Africa, these two provinces are not arid environments. Yet despite high levels of water resource availability many people are effectively excluded from access to safe convenient water supplies (Table 4.1). This is either because there is no functioning engineered supply where they live or because there is a supply, but they cannot afford to buy access to it. Such statistics are not particularly reliable, but they do show the ongoing challenge of providing supplies in this area.

Table 4.1 Contemporary Access to Piped Water in Cameroon

National population with access to safe water supplies	32–43%
National rural population with access to safe water supplies	45%
South-West rural population with access to safe water supplies	21%
National population with access to water inside their homes	11%

Sources: low national figure Government of Cameroon (1999), high national figure FAO (1994), rural figure FAO (1994), South-West figure DRA consultants for South West Development Authority (1997), figure for water in homes SNEC (Société National des Eaux du Cameroun)

Women and the Production of Water in the Pre-Colonial Period

At the beginning of the twentieth century water was already being engineered in the sense that humans were transforming their locality to suit their needs. In 1900 the vast majority of the 700,000 people (Kuczynski 1939) living in what became anglophone Cameroon collected water from springs or wells. Where such sources were unavailable the pools of fresh rain water that gathered in tree roots or leaves were used. Amongst the Bakossi people around Tombel, where the roofs of houses were made from the leaves of tree ferns, gutters were constructed from hollowed out plantain stems, which harvested rainfall into water butts (interview with S.N. Ejedepang Koge (historian of the Bakossi people), Tombel, April 1999). In some places, troughs made from tree trunks were built and put outside the home to catch the rain (interview with Fokwang Fofuleng, Bali, April 1999). At springs communities constructed small dams so as to create pools suitable for washing, and built spouts under which it was possible to place a calabash to collect water to carry back to the home. Sometimes, if distances were short, extended gutters made from plantain or bamboo were used to channel drinking water down a slope nearer to settlements. In most cases maintenance of the path that ran to the spring was an annual communal activity. Once water was carried back from the spring it was stored in earthenware containers (often on the right-hand side of the door in a corner of a principal building). Water was allowed to settle for some time in these pots so that sediment dropped to the bottom. Water at the turn of the century was both a technological and a social product.

What role did women play in the tasks of gathering water? Consider a situation where a community collected water from a nearby spring. First, women were involved in the communal work of constructing the path to the spring. The standard view from colonial documents was that it was primarily the men who undertook the labour of communal work. However, the division between the labour task itself and the associated social practices is less marked than was assumed. The labour itself is only half of the task; communal work is often followed by communal eating and drinking. So the preparation of food on communal work days was part of the pre-colonial production of water. Second, as the women were the principal carriers of water their passage maintained the path to the spring on a daily basis. Third, since it was generally women who were present at the spring, it was presumably women who maintained codes of behaviour there, perhaps only turning to men when they needed to enforce those rules. So, even before the first pipe was laid women were key to the production of water.

Women and the Production of Water in the Colonial Period

The earliest piped water supplies in the area were constructed during the German colonial period in settlements near the coast. Two were built by the colonial state between 1900 and 1903 in Buea and Limbe and two were built at missionary stations, one near the coast and one further inland near Tombel at Nyasoso. All these systems were primarily used by the colonists, with a token tap for the use of Africans. As new plantations were established in the early twentieth century so the number of piped networks increased. These water supplies were used by Africans, but that had more to do with the reproduction of labour power than any welfarist sentiment (Ardener, Ardener and Warmington 1960, Konings 1993). These supplies were constructed using forced male labour and provided water to the plantations, which were predominantly male environments (during the German period). Such limited construction work that took place was embedded within the thoroughly male world of colonial engineering.

Yet it is possible to find traces of women's involvement in the archival record. By the 1930s colonial officials were well aware that women helped to generate the cash within a domestic unit that paid for water levies for example: 'Bayangi women can sell their own products, buy their own salt and clothes and contribute towards their husbands tax' More importantly, advocates of investment in water supplies within the colonial bureaucracy incorporated women into their arguments by portraying them as the ultimate beneficiaries of improved water supplies (Buea National Archive (BNA) File Se(1933)2 'Measures for the improvement of the condition of women'). Colonial officials took the view that because the local men who dominated the native authorities (which constituted the institutional basis of indirect rule within this area) did not carry water themselves they could not see its importance. The figure of the woman as a water consumer became central to the lobbying that preceded construction. In addition, women (as mothers and housekeepers) were identified as the social group through which the European colonial state would purge Africans of their unsanitary ways. So in 1934, when money became available specifically for 'the improvement of the condition of women', opinion was divided between those who believed it should be invested in education about sanitation and home-making (primarily the missionary organisations) and those who believed it should be invested in water supply infrastructure (primarily the medical establishment and district officers).

The figure of the woman as both domestic guardian and water consumer became increasingly important in the 1950s when 'community development' was institutionalized as a means of building infrastructure in rural areas (Community Development Bulletin 3, BNA File Se(1950)2). As E.R. Chadwick, the official responsible for this development strategy, wrote:

> In some cases women have to carry water to their villages for distances of up to five miles. To do away with this drudgery so that the women may have more time to improve their homes, the people themselves in a great number of villages have constructed, under the guidance of a rural water supply officer of the Public Works Department, and with the support of the District Officer rain water storage tanks around their schools and other buildings. The people have supplied sand and stone and worked without payment. (1950s)

Women were identified as a target for community development because 'the standard of living is the standard of the home and the home is in the charge of the housewife or mother' (Chadwick 1952). So though women were not included in technical training programmes, they were to be included in village discussions (Chadwick 1950a). In addition, it was suggested by Chadwick that the project funding for community development projects could be handled by women (Chadwick 1950b). Women were even to be involved in the physical labour of construction.

> All that was necessary was for everyone to join together and for each to do a share of the work … this man to collect stones, this man to fell some trees, this man to dig, this man to fetch water … this man to do this and this man to do that. All the men in the village, and some of the women too, were given work to do. (Chadwick 1950s)

The close association between women, water and domesticity was firmly established in the minds of colonial officials. Water was seen largely as a woman's thing and women played a vital part, not only in the ideological justification for particular forms of investment, but also in the implementation process. In this respect water supplies during the colonial period were produced both for and by women.

Women, Community Development and Rural Water Supply 1960–1990

The lasting significance of colonial ideas about community development was that they produced a template for rural water supply that continues unbroken to the present. After independence the government of Cameroon swiftly ran up against financial constraints. The demands for water supplies from towns and villages rapidly exceeded the availability of capital to build them. In these circumstances community development became a useful policy: 'The main value of the organisation lies in its ability to tap and organize community self-help … there is a vast reservoir of voluntary labour which can be tapped once an organisation is established for the purpose' (BNA ExCo Memoranda 11 April 1961). Capital expenditure on water supplies by the state soon became conditional on a community being willing to donate labour, materials and some proportion of the costs to the project. A whole government ministry was formed dedicated to this approach to development. This body, often in co-operation with a Swiss NGO, organized the construction of over 100 separate rural water supply systems between 1964 and 1988 (Helvetas 1989).

These post-colonial development projects incorporated women in three main ways. First, women were involved in raising the money needed for the 'community contribution'. Rural water supplies in Cameroon are often the concretization of women's capital. Women were often the community's treasurers and they also acted as entrepreneurs looking for new ways by which the community could raise funds (Niger-Thomas 2000). Women sometimes established new communal farms, the profits from which were put towards water supplies. They monopolized and co-ordinated sales of palm wine in local markets for the same purpose. Since water was identified as of particular benefit to women there was also a degree of solidarity between those women who had moved from the village to town and prospered (external elites) and the women who had remained in the village. If the external women's groups wanted to support the women in the village they often chose to finance water supplies. Men also made financial contributions. Indeed the formal individual levies for men were usually twice those of the women, but women were often more zealous in raising this money, and had to be more imaginative in doing so since they tended to have less ready access to a cash income.

Second, women were also often active participants in the physical construction of the projects. In the vast majority of cases the supplies were gravity-fed piped systems from spring catchments. There was considerable variation in the degree to which women were involved with labour in different places, but some generalizations can be drawn. Women were considered capable of carrying materials and digging trenches, but not of cracking stones, working with concrete or undertaking building work. This division of labour reflects both local cultural values and the values of the different external institutions which were involved in the projects. Universally, however, women were expected to prepare the food for the occasions that marked key rites of passage during the project cycle: project initiation meetings, visits from donors, completion of major elements and project inaugurations. The women's work of supplying and preparing this food should be incorporated into the cost of production.

Third, women were producers of these projects in as far as they were the people who demanded that water supply systems should be built in preference to other projects. For example in Tombel, the small town mentioned at the start, it was the women who asserted the importance of water at significant public meetings. A piped water project had first been proposed in the 1930s but for a variety of reasons was never built. In 1959, Cameroon held elections for prime minister, and Tombel was one stop on the campaign tour of John Ngu Foncha, the eventual victor. At a mass rally he demanded to know what the people of Tombel needed. One woman, who was not an activist from his political party, stood up and told Foncha that the women of Tombel would like to vote for him, but they were ashamed to go to vote because they were dirty. They were dirty because there was no water for them to wash with once they came back from their farms. To demonstrate that she was dirty the woman took off her dress and stood before the crowd naked, and continued her complaint by saying that they would have liked to prepare a meal for Foncha, but that they were ashamed because they had nothing to boil the food in except urine.

This melodramatic episode can in part be understood as a party political challenge and in part as a gender political challenge. The woman was taking a well-established form of protest against the behaviour of specific men and adapting it to the circumstances of party politics (Ardener 1996[1970], Awasom 2002). Earlier protests had used nudity, faeces and urine to play on male anxieties about impotence in order to humiliate individuals. As Shirley Ardener puts it, 'when the women of Cameroon subject a man to such a display they demonstrate that they no longer recognize his power to elicit conformity' (1975: 43). It is unclear from local ethnography (Balz 1995, Ejedepang-Koge 1971) whether such forms of protest were a feature of the local Bakossi society or whether they were imported from the North-West province by migrant labourers. However, the meaning was clearly understood: the language of insult was being used to provoke the politician to deliver on his claim to be powerful. Foncha subsequently took a personal interest in the case and lobbied hard for the construction of the Tombel water supply.

Women and Water in Contemporary Cameroon

Many of the ways in which women have been involved in the production of water continue to the present, but it is their role in protests that is particularly significant. Where men often chose to register their complaints about water rates through petitions to government or letters to the newspapers, women were more inclined to take direct action. These protests generally relate to two state policies. First, the closure of public taps and, second, the attempts to charge for water at public taps. Both came to the fore in the 1980s.

Water rates were first legally introduced in 1934, but for many years they only impinged upon the lives of a tiny proportion of the African population because improved water supplies were so limited. The official policy of the British administration was that those who used piped water ought to pay for it in proportion to the amount they used. The administration also aspired to set water rates at

Table 4.2 Rising Levels of Subsidy in Cameroon's Water Supply Sector

	Income	Expenditure
1935–1936	£387/19/1	£257/2/1
1937–1938	£411/7/-	£423/9/10
1950–1951	£404	£1,770
1954–1955	£753	£1,635
1961–1962	£11,214	£18,439
1971–1972	4.4 million CFAF	125 million CFAF
1979–1980	21 million CFAF	Unknown
1982–1983	12,300 million CFAF	12,900 million CFAF
1985–1986	16,500 million CFAF	19,400 million CFAF

Notes: after 1980 the figures refer to all of Cameroon as opposed to just the two anglophone provinces. CFAF = Communauté Financière Africaine Franc
Source: Page 2000

a level that generated sufficient income to cover the running costs. However, the difficulty of collecting rates meant that water supplies were subsidized from general revenue. After the expansion of the number of urban networks in the late 1950s many more people were expected to pay for water. To make it simple to gather, the annual water levy was included in the poll tax collected by local councils. From the perspective of the central government this had the advantage that many people were unaware of the fact that they were paying for water, but it had the disadvantage that much of the money was siphoned off by local councils. Even if all the water bills had been collected the rate was set at such a low level until 1980 that it did not cover the running costs of the systems (Table 4.2).

Between 1960 and 1972 attempts to raise the water rate were blocked by the House of Chiefs, the second chamber of the parliament, which objected to the idea that water could be sold. As a result this pattern of an increasing deficit continued until the administration of water supplies was transferred from a government department to a quasi-autonomous national corporation called SNEC in the early 1980s. This transfer coincided with a huge investment in infrastructure, during which most of the water supplies now operating throughout Cameroon were constructed. The planning for this expansion had been carried out in conjunction with UNICEF and the WHO and loan funding came in large part from the World Bank. The systems were generally built by international engineering consultancies without community involvement. During this rebuilding, universal chemical water treatment was introduced and electric pumps were brought into widespread use, which made systems safer and more reliable but also more expensive to operate. The experience of the move from local government to national parastatal was mixed for most households in Cameroon. On the one hand outdated technology was replaced by more regular, higher quality water, but on the other hand the water changed in taste and price (Table 4.3). After 1980, all new house connections and public taps had water meters. Householders with water inside their homes and local councils with public taps started receiving monthly bills. Regardless of the improvements in infrastructure the new water supplier rapidly became extremely unpopular as a result of its determination to charge an economic price for water.

Councils found that the cost of supplying water to public taps was eating up between a quarter and a third of their annual budgets. As a result many started to default on their water bills. In 1998 for example, local newspapers reported that the council in Buea owed SNEC 520 million CFAF and the council in Tombel owed 113 million CFAF. SNEC treated public taps in the same way they did private account holders; those councils who failed to pay their bills had their public taps disconnected. Towns such as Bali, which regularly defaulted on payment, rarely had more than a few public taps that were operational. Indeed the council there never persuaded SNEC to open more than nine of the thirty-six taps that were built during the 1984 reconstruction.

The struggle to keep public taps open has been bitter and protracted. The reduced number of public taps resulted in long queues to collect water and increasing frustration. From the perspective of SNEC and the government of Cameroon such

Table 4.3   The National Water Price Rises 1975–1999

	Price of domestic water (CFAF/m^3)	French Franc Equivalent (FF/m^3)
1963–1970	7.7	0.154
1971	30–70.5	0.6–1.4
1982 (January)	125	2.58
1982 (April)	222	4.5
1986	244	4.88
1999	271	2.71

Note: the CFAF was devalued in 1994
Sources: 1963–1970 figure: BNA Rd(1961)2, Holloway Report BNA Rd(1961)3, and Efungani personal papers. 1971 figure: BNA Rd (1965)2. Other figures: *Cameroon Tribune* 13 January 1982, *Cameroon Tribune* 18 August 1982, *Cameroon Tribune*, 10 October 1986

inadequate public supplies were an incentive for households to bring water into the private sphere. But from the perspective of many households the cost of installing a domestic water supply was completely prohibitive. Only since the rapid decline in standards of living in the mid 1990s have the authorities recognized that public taps are set to remain part of the urban landscape for the foreseeable future.

The battle to preserve public taps and access to 'free' water has been one in which women have taken a central role. For example, in Limbe in August 1982, 200 women marched on the office of the Senior Divisional Officer, the most important government official in town.

Carrying plastic and enamel buckets, the women chanted slogans denouncing the activities of SNEC and calling for a complete overhaul of the water system in Limbe urban township ... Prominent amongst their complaints was the inconsistency of the water supply in Limbe town in general and New Town in particular. The women also complained against the spiralling water bills with little or no water flowing from the taps. Finally the reduction in the number of public stand taps meant increasing hardship for those without water in the house... The Senior Divisional Officer thanked the women for the orderly manner in which they presented the grievances and promised to look into the matter. He appealed to the women to be patient since matters of the sort could not be redressed overnight. Apparently satisfied with the Senior Divisional Officer's promise, the women trooped back to town chanting 'Massa we thank you for your promise.' (*Cameroon Outlook*, 26 August 1982)

The officials were surprised by the protest which was unusual during a time in which sycophancy dominated public language and repression of dissent was the norm. The leaders were subsequently detained, but stuck to their story that the protest was spontaneous and genuine.

The policy of reducing the number of public taps did not, however, change. The women were assured that the plans to upgrade the infrastructure in Limbe were in hand, but at the same time SNEC announced that the number of public taps would be reduced from seventy-one to forty-two (*Cameroon Tribune*, 24 April 1985). Once the construction work actually started in 1985 the central government Ministry revealed that the number of public taps that they would pay for had been reduced to twenty-one. The local council were given the option of paying for additional taps, but at the specified cost of 560,000 CFAF/tap they could not afford to do so. However, since the women forced the authorities to recognize that many households in Limbe could not afford to bring water into their homes the policy has changed. SNEC now acknowledge that such communal taps perform a vital social function and have abandoned their policy of removing them entirely.

Having recognized the need for communal public taps, SNEC have looked at other ways of making them financially sustainable. In particular, since 1993 SNEC have been trying to privatize the public taps. This entails closing the tap, then leasing it to an individual, who will stay by the tap and sell water at a recommended rate of 5 CFAF for a 10 litre bucket. SNEC claim that this rate should generate a sufficient profit over a month to make this an attractive business proposition, without denying anyone access to water. Since the individual who takes on the lease is charged for all the water distributed from the tap they have a financial incentive to ensure that none is wasted and they will report damage as soon as it happens. This system was introduced to a number of towns across Cameroon in the 1990s. However, in Limbe the threat of the privatisation of public taps generated sufficient anxiety to bring the women onto the streets once more. Even more dramatic, however, was the opposition to such a policy from women in Tombel.

In 1993 the Governor of the South-West Province announced the new policy of privatising public taps and told the Tombel people that they should be prepared to start buying water from the taps. The people immediately refused and in response more public taps were closed by SNEC. In January 1994 the Tombel Women's Association wrote to the local Divisional Officer complaining that they could not afford to pay for water, and that they resented paying for a service which was declining in quality. They demanded that SNEC should leave Tombel and that the treatment works constructed in the early 1980s should be disconnected and replaced by the system constructed in 1963. Though the newer system was more elaborate and had a physical and chemical water treatment plant it was associated with SNEC. The women preferred the earlier, simpler network which was not seen as belonging to SNEC or indeed to the government, but was generally perceived to have been a gift from Prime Minister Foncha to the people of Tombel, and in particular to the women of Tombel. After a week when nothing was done, they went as a group to the old reservoir and began clearing the vegetation around it as a means of expressing their desire that it should be reconnected.

After a further fortnight of inactivity 4,000 women marched through town to the SNEC office, which was besieged, and various 'traditional' rites were carried out. At the front of the crowd some old women marched naked, and on reaching the SNEC office they urinated on the steps. As one witness to events put it 'nakedness is the highest and last stage of the women's society in Bakossi. From there only death can solve the problem' (interview Mr Epote, Buea, 20 April 1999). The Tombel men were in hiding, for fear of seeing these old women naked (interview with Mr Sone, Buea, 21 April 1999). Various herbs were thrown down on the steps of the office, which were said to be able to transform into poisonous snakes if the threshold was crossed again (interview with S.N. Ejedepang-Koge, Tombel, 26 April 1999). The SNEC employees who had been warned of the gathering crowd had already fled and refused to return to the office 'partly to avoid the wrath of the people and also to keep away from the "juju" put there by the women' (*Herald*, 14 February 1994).

The protest was co-ordinated by a woman who had spent many years working for Community Development and had been actively politicized as an advocate of women's rights. As with the earlier use of nudity in 1959, this was a reworking of elements of traditional protest in a modern context. The engineers who ran the water in Tombel, the Divisional Officer, the Governor of the Province, the local Member of Parliament and the traditional authority (chief) are all men. The form of the protest adopted a template from gender conflict because the forces who appeared to be conspiring to remove free access to water were men. Of course, the majority of men in the town also objected to paying for water and the protest cannot be read solely in gender terms. Since SNEC is a parastatal corporation an attack on SNEC is an attack on the government. Since the government emphasizes its respect for 'tradition' it is only by using 'traditional' forms of protest that communities are able to challenge the state in a context where state violence is always a possibility. Local government bureaucrats in Tombel have been able to negotiate a path in which they both condemn the expulsion of a government agency but also express their respect for the women and their assertion of their traditional rights. Local male leaders have been able to preserve good relationships with the governing political party by categorizing the action as 'traditional' and therefore in some ways tolerable. So the protest was not only a protest against men, it borrowed forms of gender protest and reinterpreted them in a way which suited the contemporary political context.

After expelling SNEC from town the women handed the water system over to a community committee, dominated by men. The SNEC treatment works was abandoned, but the SNEC distribution network is still used. A small number of public taps have been kept open and there is no charge for their use, though yard taps and private connections incur a fee. After dissatisfaction with the probity of the individuals on the water committee, the key post was given to the woman who had been involved in the initial organization of the 1994 protests. Under her leadership the committee managed to break even and in fact they were even able to save enough money to plan some small extensions to the system. However, political grievances and ambitions within the Tombel community have meant that she has been sidelined and the future of the water committee is now less clear.

Conclusions

The story of women's contribution to the production of water in Cameroon in the twentieth century shows at least two things. First, the idea that women are only connected to questions of water policy and water history because they are water consumers is inadequate. Rather than start our discussions about women and water at the point where the women collect the water from a well, put it on their heads and set off home it is necessary to ask: how did the well get there in the first place? And, what role did women take in the process of bringing that well into being? This is more than the usual gripe that writing around development policy and development projects tends to have a casual disregard for history and the particularities of place. Rather the logic of 'women in development' rests, ironically, on a denial of the role women have played in the past. That argument runs something like this: the rate of failure of water projects (urban and rural) is high because women have been excluded from the project cycle in the past. Including women, it is suggested, can transform failure into success. To sustain this argument it is necessary to emphasize the difference between past and present in terms of women's participation. Such strategies seem historically untenable in Cameroon, they deploy a simplistic formulation of power and they rest on the assumption that women have played no part in the making of particular places. The empirical material presented here aims to contest such a view. This is not to say that more determined efforts to include women in project planning, implementation and evaluation are not merited; there is undoubtedly scope to include women in projects in increasingly meaningful and innovative ways. But it is important to remember that women in Cameroon have been central to the rural development discourse for at least thirty years, and have contributed to the making of rules of water management and the reproduction of the cultural values around water for far longer. There is no need to deny the asymmetries of power between men and women in order to admit the possibility that women have, in a variety of ways, influenced the development of Cameroonian water supplies over the last century.

It might be claimed that the particular case study described is an unusual one and that it is not possible to generalize about the role of women in the history of water supplies. But, at the very least, this case ought to raise the question of whether similar narratives can be considered in other places. To ignore the institutions and historical triumphs that already exist may be to miss an opportunity to instrumentalize a potentially progressive tradition of gender activism. Rather than incorporating women into an arbitrary, uniform and externally conceived project cycle the best opportunity for empowering women may be to publicize their historic achievements in places like Tombel. In this town women have challenged attempts to close down public taps and make water-users pay at the point where they collect their water; they have been at the forefront of public hostility to the idea that water is best treated as an economic commodity. As part of the resistance to these policies the current generation of women have emphasised the role that their mothers and grandmothers played in the construction of water supplies in the 1960s. Their objective has been to defend existing rights of access. Water has become a symbol of the benevolence of the first generation of independence leaders and the competence of communities in relation to the inept government and its grasping water corporation. Women have not been alone in calling on recent episodes in the history of water development to help justify action against further privatization of drinking water in Cameroon, but their actions in the past have often been unfairly neglected.

Second, it is necessary to expand what is understood by the 'production of water' in order to reveal the crucial role women have played in the past. The construction of water supplies not only includes the actual direct labour of moving materials, digging pipelines and building (in which women have at various times participated) but the ancillary labour such as preparing food for community labourers and indirect labour such as collecting money or running committees (through which a community's wealth has been transformed into pipes, cement and technical expertise). It is necessary to extend the notion of production to include the making of ideas as well as the making of infrastructure. This includes both the symbolic meanings that accrue to water and the ideas about the correct (most efficient, most just) way to deliver water to those who use it.

In recent years social sciences have been enchanted by 'consumption' as a neglected sphere of research whose importance was undervalued because of an earlier obsessive prioritization of 'production'. The use of the idea of the 'production of water' as set out here is not a bid to reverse the current trend so much as to start bridging a false dichotomy. Consumption can only be understood in relation to production and vice versa. It is precisely because of women's involvement as water consumers that they were involved in the production of water in the past; and it is that past involvement which may underpin their present claims to rights of ownership and access.

References

Appadurai, A. (1986), *The Social Life of Things*, Cambridge: Cambridge University Press.

Ardener, E. (1996 [1970]), 'Witchcraft, Economics and the Continuity of Belief', in S. Ardener, (ed.), *Kingdom on Mount Cameroon*, Oxford: Berghahn.

Ardener, E., S. Ardener and W. Warmington (1960), *Plantation and Village in the Cameroons*, Oxford: Oxford University Press.

Ardener, S. (1975), 'Sexual Insult and Female Militancy', in S. Ardener, (ed.), *Perceiving Women*, London: Dent.

Awasom, S. (2002), A Critical Survey of the Resuscitation, Activation and Adaptation of Traditional African Female Political Institutions to the Exigencies of Modern Politics in the 1990s: The Case of the Takumbeng Female Society in Cameroon', Paper presented to CODESRIA 10th General Assembly, Kampala, Uganda, 8–12 December 2002.

Balz, H. (1995), *Where the Faith Has to Live: Studies in Bakossi Society and Religion*, Berlin: Deitrich Reimer.

Castree, N. (1995), 'The Nature of Produced Nature: Materiality and Knowledge Construction in Marxism', *Antipode*, 27(1): 12–48.

Chadwick E.R. (1950s), 'Community Betterment in Africa', Pamphlet published in the early 1950s by HMSO. BNA loose in File No. Se(1950)2.

—— (1950a), 'Community Development Bulletin No. 13', June 1952 and 'Community Development Bulletin No. 14' July 1952, Circulars BNA File No. Se(1950)2.

—— (1950b), 'Community Development Bulletin No. 2', Circular April? 1950, BNA File No. Se(1950)2.

—— (1952), 'A short list of quotations concerning Community Development', Pamphlet 8 May 1952, BNA loose in File No. Se(1950)2.

Diduk, S. (1989), 'Women's Agricultural Production and Political Action', *Africa*, 59(3): 338–55.

—— (1997), '"The Only Men Left in the Land are Women": Rural Women's Protests in the Republic of Cameroon, 1990–6', Paper presented to Centre for Cross-Cultural Research on Women, Queen Elizabeth House, Oxford University, 5 June 1997.

Ejedepang-Koge, S. (1971), *The Tradition of a People: Bakossi*, Yaounde: SPOECAM.

Endeley, J. (2001), 'Conceptualising Women's Empowerment in Societies in Cameroon: How Does Money Fit in?', *Gender and Development*, 9(1): 34–41.

Fonchingong, C. (1999), 'A Difficult Route: Structural Adjustment, Women and Agriculture in Cameroon', *Gender and Development*, 7(3): 73–79.

Forje, C. (1998), 'Economic Crisis Helps to "Demarginalize" Women', *Development in Practice*, 8(2): 211–16.

Goheen, M. (1996), *Men Own the Fields, Women Own the Crops: Gender and Power in the Cameroon Highlands*, Madison: University of Wisconsin Press.

Harvey, D. (1974), 'Population, Resources, and the Ideology of Science', *Economic geography*, 50: 256–77.

—— (1993), 'The Nature of Environment: The Dialectics of Social and Environmental Change', in R. Miliband and L. Panitch, eds, *Socialist Register*, 1–51.

Helvetas (1989), *Evaluation of Water Supply Systems Constructed from 1964 to 1989*, Bamenda: Helvetas.

Konings, P. (1993), *Labour Resistance in Cameroon*, London: James Currey.

Kopytoff, I. (1986), 'The Cultural Biography of Things; Commoditization as Process', in A. Appadurai, ed., *The Social Life of Things,* Cambridge: Cambridge University Press.

Kuczynski, R. (1939), *The Cameroons and Togoland: A Demographic Study*, Oxford: Oxford University Press.

Niger-Thomas, M. (2000), *Buying Futures: The Upsurge of Female Entrepreneurship; Crossing the Formal and Informal Divide in Southwest Cameroon*, Leiden: CNWS Publishers.

Page, B. (2000), A Priceless Commodity: The Production of Water in Anglophone Cameroon 1916–1999, Unpublished D.Phil thesis, University of Oxford.

—— (2003), 'Communities as the Agents of Commodification: The Kumbo Water Authority in Northwest Cameroon', *Geoforum*, 34: 483–98.

Smith, N. (1984), *Uneven Development: Nature, Capital and the Production of Space*, Oxford: Blackwell.

Strang. V. (2004), *The Meaning of Water*, Oxford: Berg.

Swyngedouw, E. (1997), 'Power, Nature, and the City. The Conquest of Water and the Political Ecology of urbanization in Guayaquil, Ecuador: 1980–1990', *Environment and Planning A*, 29: 311–22.

Swyngedouw, E., M. Kaika, and E. Castro (2002), 'Urban Water: A Political-Ecology Perspective', *Built Environment*, 28(2): 124–37.

Uitto, J. and A. Biswas (2000), *Water for Urban Areas: Challenges and Perspectives*, Tokyo: United Nations University Press.

Van Wijk-Sijbesma, C. (1998), *Gender in Water Resources Management, Water Supply and Sanitation: Roles and Realities Revisited*, Technical Paper series no. 33-E (updated version of technical paper 22), The Hague: IRC Publications.

5

Geology and Gender: Water Supplies, Ethnicity and Livelihoods in Central Sudan

Anne Coles

Introduction

This is a historical study of 200 villages, their water environments, people, livelihoods and settlement patterns. It relates the physical factors that define the potentially available water to the human factors affecting this water's exploitation and use. It explores the complex interrelationship between water supplies, gender, ethnicity and ways of life, which ranged from permanent settlement to nomadism.

In the late 1950s and early 1960s there was considerable optimism about the development of water resources in Africa, mostly focussed on large-scale dam-building. Much less academic attention was given to changes resulting from more modest improvements in water supplies for rural communities, such as form the subject of this chapter.

The area studied lay in east-central Sudan between the Blue Nile and the River Atbara, close to the Ethiopian foothills, and formed part of Gedaref District. It was part of the vast savannah belt that stretches across sub-Saharan Africa from west to east, a semi-arid area where the constraints of the physical environment were and are very real. Gedaref had been an area of immigration since the 1890s. Local Arabs, incoming Western Sudanese and West African migrants found different combinations of the area's very varied water sources appropriate to their ways of life. But while modern technology enabled greater use to be made of the land, it was less successful in creating water supplies for new, stable villages.

By chance, the research was carried out in a year of drought (1961). It illustrates the continuing marginal adequacy, and erratic character, of many water supplies in the African savannah, especially those dependent on annual rainfall for their replenishment. It shows the social and economic disruption to men, women and communities, and the inevitable reduction in families' living standards, caused by unexpected water shortages.

The research relates to a time before gender analysis became an accepted part of academic fieldwork. Yet an appreciation of gender relations was essential to understanding how households perceived and made use of water resources. The very different roles played by men and women in domestic water acquisition and management, and variations in family composition, were integral to the study. Cultural gender norms, partly mediated through the fetching of water in the dry season, influenced both economic opportunities and the choice of sites for settlement. Gender relations varied according to ethnicity, with its connotations of culture, nationality and power, as well as in terms of aspirations and religiosity.

The Area and its Physical Characteristics

The dominant physical feature of the area was the clay plain, which covered large areas of central Sudan and here sloped gently down to the River Rahad in the west. Its monotony was broken by two groups of hills. In the middle of the area, around the small town of Qala' en Nahl, there were isolated clusters of granite domes and serpentine hills, outcrops which formed part of Africa's ancient basement complex rocks. In the east, the gentle slopes of Gedaref's basalt ridge overlay Nubian sandstone and were the site of the district headquarters, Gedaref Town.

The average rainfall ranged from about 480 mm a year in the NNW to about 700 mm a year in the SSE. Importantly, rainfall was highly variable. In the previous fifty years, rainfall at Gedaref Town had varied from 64 to 160 per cent of the mean. Moreover, variability increased as average annual rainfall decreased – the less the rainfall, the greater the fluctuations. The date when the first serious rains occurred was also highly variable. This variability had a marked effect, both on water supplies and through them on ways of life and livelihoods.

The natural vegetation generally reflected the rainfall. It ranged from short annual grasses in the north to open deciduous woodland with tall perennial grasses in the south. Average rainfall was everywhere adequate for cultivation – sorghum being the main crop. But reliance on one crop was risky. Livestock rearing formed an insurance against crop failure, especially in the north, while in the south sesame, which matures earlier than sorghum but requires more moisture, was widely grown as a second crop. Traditionally, however, large areas of the clay plain were under-used for lack of drinking water both for people and for livestock.

Water Environments for Settlement

The different water environments attracted rather different peoples and produced rather different settlement patterns (see Tables 5.1 and 5.3). The River Rahad, which rose in the Ethiopian Highlands, provided water throughout the year. Although it flowed seasonally, ample water was always available from pools or from shallow excavations in its bed. It provided water for almost 30 per cent of the villages studied, a string of permanent settlements along its course. (The south, being wetter, was unhealthier in the rainy season.) Additionally the Rahad was the site of the dry season camps of those living in the Qala' en Nahl Hills who became transhumant. The river banks were also used by Arab nomads from the Butana desert, who moved long distances southwards with their stock when northern pastures dried up, returning once the rains produced fresh grass in the semi-desert.

The most important hill mass as far as water was concerned was the Gedaref Ridge. Around 40 per cent of all villages were sited here and wells sunk into the basalt rock with an average depth of 13 metres were their main source of supply. The best and shallowest were near the centre of the Ridge where run-off from relatively large catchment areas was concentrated in short streams. Gedaref Town's wells were of this type. The deepest and lowest-yielding wells – those over 25 metres seldom supplied useable quantities of water – were on the margins of the Ridge. People living here often made local movements back to the centre to fetch water from its better wells in the dry season. Well water in the basalt frequently became salty in the dry season but, despite total dissolved solids and nitrates well above conventional potable limits, such wells continued to be used because of lack of an alternative source. Where, occasionally, Nubian sandstone outcropped, wells averaged 25 metres.

The hills of Qala' en Nahl had several rather inadequate sources of water. Unsurprisingly barely 20 per cent of villages were sited there. Wells in the granite, sited in the weathered pediment zones below the rocky hills, had an average depth of 18 metres but, with their limited catchment areas, they had very marked seasonal changes in water level and water availability. Traditionally many villages had to practice transhumance. But the granite provided two additional minor sources of water that were important. *Jamams* were shallow excavations in the angle of a hill foot where water seeped sufficiently throughout the dry season to support village caretakers. *Gallits* were rock pools which filled with the first rains and enabled men to return to clear fields ahead of the main time for sowing. Moreover, hill-foot villages throughout the area but particularly in Qala' en Nahl, often had horse-shoe shaped, hand-dug ponds or *hafirs*, sited close to the hills to receive direct run-off. These held water for a few months after the rains, depending on the number of animals and sometimes people using them. Wells in the serpentine provided water throughout the year (and a pumped supply supported the town of Qala'-en-Nahl) but, with an average depth of 40 metres, they were the deepest in the area.

Since the late 1940s there had been two important developments by government. Mechanical excavators had been used to dig larger, deeper *hafirs* with feeder canals with an average capacity of 15,000 cubic metres and these could be located further out in the clay plains. There were over fifty successful ones in the area. Some were constructed to enable mechanized sorghum production on a large scale in areas designated as Mechanized Crop Production Schemes (MCPSs). Some were sited to enable established villages to become permanently settled. Others were in still-underused areas of 'bush'. Many, particularly those in the virgin lands, had attracted immigrants. Additionally bores had been drilled in the Nubian sandstone, though these had been disappointing – only a few still produced water three years after construction. Altogether almost 15 per cent of villages were now situated in the clay plains, relying on mechanically excavated *hafirs* (or occasionally bores) as their main source of water supply. Importantly, all water sources, except for the River Rahad and to some extent the bores, were dependent on replenishment from the area's annual rainfall.

History: The Development of Water Resources

By the eighteenth century there were certainly settlements based around wells in basalt and granite rocks, with, additionally, hand-dug *hafirs*. Ottoman rule from Egypt from the 1820s onwards brought improved techniques for accessing water supplies. In the 1880s a revolt by the Mahdiya movement devastated and depopulated Gedaref, but the Anglo-Egyptian (predominantly British) condominium from 1899 onwards enforced peace. Pilgrimage increased – Gedaref lay on the overland route to Mecca – and by 1912 pilgrims (there were almost as many women as men) from as far as Senegal and Gambia were established in the area (Mather 1954). The completion of the railway between Sennar and Port Sudan in 1926 linked Gedaref with the outside world. The population, now racially very mixed, essentially settled in traditional areas where water supplies were readily available. Villages were clustered at the foot of hill masses or strung out along the River Rahad.

Responding to the increasing immigration, the administration developed new water supplies. Modern techniques were used to bore wells in the serpentine areas, opening up a new water environment. Later, in 1920–30, over forty *hafirs* were initiated by the government to relieve pressure on existing water sources resulting from population growth.

The Area's Ethnic Diversity

Gedaref was an area of economic opportunity. There were three main population groups of very roughly equal size.[1] First, the indigenous Arabs. Then people from Western Sudan, both genuinely settled families and young single men who came for a few years to gain sufficient money to establish households back home. A minority, mostly lone men or childless couples, became perpetual wanderers, cultivating here or labouring there. Third, there were many West African foreigners, some still on their way to Mecca, pausing to earn cash to complete the journey, others having settled permanently on their return. Within these broad groupings over 150 tribes were represented.

While some of the Western Sudanese were from Arab tribes, many had a culture that was essentially African and only superficially Moslem. Among the West Africans, many Chadians had more in common with neighbours in Darfur than with those from Nigeria. Ultimately, social and economic characteristics tended to be modified to conform to local Arab norms, depending both on the immigrant's length of time in the area and intentions to stay. For example, a few old-established Hausa villages, despite lacking a stock-rearing tradition, had invested in cattle, which followed Arab nomadic grazing patterns. Because land was available, almost all immigrants became, like the Arabs, cultivators in their own right, although some Western Sudanese single men preferred to work as labourers. Almost all immigrants adapted to growing *dura* (sorghum) for their staple diet because *dukhn* (pennesitum millet) yielded poorly on Gedaref's clay soil. Incoming men rapidly acquired Arabic, the area's dominant language. Women, much less involved in life beyond the home, lagged behind in learning Arabic, which only gradually gained ground as the language spoken in the family. Of particular interest was the extent to which villages had mixed populations, even though individual tribal or regional groups might be spatially separate within them. While some villages consisted of a single tribe or a single group, about a quarter contained people from two or more groups.

Gender and Family

Throughout Sudan the basic social and economic unit was the extended family. Most of the more settled people of Gedaref lived within this framework with its mutual rights and obligations and male dominance. Women were responsible for managing the home and for the care of dependants. Society was at its most patriarchal in the old Arab villages, where male heads of the most prominent families often assumed responsibility for all those within their compounds, including adult sons, their families and retainers, although the average family size in the Arab villages studied was six to seven. Among new immigrants, family units tended to be smaller. With a high regard for family life, nearly all long-settled West Africans had many children. Among Western Sudanese couples, some were childless – seemingly partly from choice, allowing women to work in the fields, and partly the result of venereal disease. Taking two new *hafir* settlements both with pioneer populations, the West Africans had an average family size of 4.6 and the Western Sudanese of only 2.8.

In all groups women occupied an inferior position, following the locally accepted interpretation of Islam. Seclusion of women was widely recognised as an ideal but one not commonly attainable. Nevertheless, better-established West African villages were usually distinguished by tall grass walls surrounding their compounds. Most West Africans were reluctant to let (or admit to letting) their wives work in the fields. But the women were not economically unproductive. In mixed villages they were widely employed as domestic helpers by Arab families, particularly doing laundry, pounding grain and sometimes fetching water. Among local Arabs, a woman might occasionally farm in her own right. In most Western Sudanese communities women were active in the fields; they generally had more independence, making a more visible contribution to the household economy, including selling prepared food in Gedaref Town market. In some cases such women seemed to be entirely self-supporting: farming independently, brewing and engaging in prostitution.

Gender Roles: Fetching Water

In much of the developing world women are the main fetchers of water for domestic use, but this was only partly true in Gedaref. Where a water point was within or close to the village, it was usually women and girls who collected and transported water. Western Sudanese firmly regarded collecting water as women's work. Arabs took a more pragmatic view: while women normally fetched water, the task might be shared if one partner was busier or much was needed. Many West Africans and strict Moslems were reluctant for young wives to undertake the task (since wells were public places) but girls and older women did so. Generally the organization and collaboration involved in drawing from a village well were a matter for women, who sometimes worked in pairs to haul the buckets where the well was deep, and ensured that access was fair if there was crowding and that young girls were helped. Village wells provided women with an important opportunity for social networking with women from their own tribe and also, often, with women of different ethnic origins, depending on the configuration of the village.

Where water was hard to obtain, men from all the groups collected it. If a village well was more than 30 metres deep, considerable physical effort was needed to raise water, so men usually undertook the task. Similarly, women did not usually fetch water from more than about half a mile away. Apart from the distance, fetching water from a neighbouring village posed social problems, for the women would be seen by strange men.

Thus when water had to be fetched from further off, a common occurrence in the dry season – for only 65 per cent of villages had water supplies that were adequate throughout the year – men undertook the task. Animal transport was increasingly essential as distance to the water point increased. Donkeys and camels were used where possible – boys were sometimes entrusted with donkeys. Where the distance was greater than about 3 miles, possession of sufficient beasts of burden determined whether a village became transhumant. Where a village was more than about six or seven miles from its dry season water point, it effectively had to be transhumant unless it was wealthy and prominent enough to receive water by truck. The time spent fetching water was closely related to the amounts needed, the containers used and the distance involved, as well as the amount of crowding around the water point. Girls usually fetched water in buckets containing roughly 9 litres. A 4-gallon paraffin tin, as carried by women (and men), held about 16 litres. However, a donkey could manage up to 77 litres using a *kurrug* (a leather or canvas pannier) and a camel over 200 litres depending on the containers used. The fact that a donkey could carry so much more than a woman was one reason why water consumption did not always decline if, seasonally, water had to be fetched from further off.

Women's Management of Domestic Water Supplies: Economy in Scarcity

At the household level, women were the users and managers of water par excellence. Water was required for drinking, food preparation, bathing, laundry and domestic livestock. Drinking water was cooled by evaporation, either by hanging it in a goatskin *girba* in a shady, breezy place outside or by putting it in a porous pot inside. A good deal of water was needed for cooking and women might vary the method of food preparation according to the relative availability of water and grain: for example *kisra* (a fermented flat pancake) used more water than *asida* (a thick porridge). Women were responsible for providing water for chickens and, often, for watering the goat(s) which provided milk essential for morning tea. Water for washing at meal and prayer times was poured from a fine-spouted jug. If necessary, men washed at the water point but women and children bathed at home. Clothes washing was also usually done at home. The same deep tray used for laundry was also used for washing small children, who would sit in it, while older ones would be scrubbed and then splashed as they stood on the ground. Crawling infants and toddlers were particularly difficult to keep clean. Maximum hygiene with minimum water was achieved by washing faces first and then moving down their bodies.

The amount of water used by individual families was studied in seven villages, which had varying ethnic and household characteristics and experienced different degrees of water shortage in the dry season. Average per capita water consumption was one 'tin' or 16 litres per day,[2] the range for six of the villages surveyed was 15–18 litres. (Average consumption in the seventh declined to 13 litres seasonally, when water was 6 miles distant. Generally, pioneer settlers and agricultural workers, typically single men, used rather less. West Africans and Western Sudanese tended to use slightly less than Arabs, who considered them dirty. Water consumption appeared related to family structure; the presence of women and children tended to raise living standards and established households often had more domestic animals, which might be watered at home. Certainly, comparison of well-established villages with ethnically similar new *hafir* villages showed that the older communities used more water than the new, even though the latter had ample, easily accessed water available.

The demand for water was surprisingly inelastic, perhaps because household tasks were already honed to be economical. It was unusual to find families where consumption exceeded one-and-a-half tins (24 litres) per head and rare to find ones where water use was less than two-thirds of the average. These figures for water use are quite high compared with international literature and later figures from Western Sudan but were comparable to government estimates (Robertson 1950). Consumption did not seem to increase when ample water was available; perhaps the sheer effort of fetching it or of paying others to do so deterred extravagance. Conversely, it seemed that only during a brief, unexpected period of shortage were people prepared to make additional economies, largely by disregarding standards of hygiene.

The Links Between Ethnic Aspirations, Perceptions of Water Adequacy and Choice of Settlement Types

Table 5.1 Distribution of Tribal Groups in the Area

	% of villages in each environment containing		
	Local Arabs	Western Sudanese	West Africans
Ridge North	77	17	23
Ridge South	24	64	31
Hills	27	75	27
River	25	16	81
Plains	9	87	25

Source: Graham 1963: 141

Different groups predominated in different parts of the area (Table 5.1). Early on under the condominium government, Arab returnees of the Shukriya tribe were given control of Gedaref North district, while Western Sudanese, who had first moved into the area during the Mahdiya, were in due course allocated Gedaref South and Qala' en Nahl councils. West African pilgrims, although foreigners, were generally welcomed because they were hard-working and pious, land was available and they were regarded as transient. At the micro-level, newcomers tended to be attracted to villages where relatives, members of their own tribe or others from their home area had settled. But there were other reasons for the settlement pattern that emerged. The three groups of people tended to have different economic aspirations, social values and ways of life and therefore had varied views of what constituted suitable sources of water and appropriate settlement sites. These differences were particularly obvious in the dry season when occupations and perceptions of water adequacy were largely determined by ethnicity and duration of settlement in the area.

Local Arabs, with deep roots in the area, were generally reluctant to abandon their traditional villages and grazing lands, even if water supplies became very inadequate. This made them more ready to accept transhumance, particularly as it enabled them to remain with their livestock for more of the year. Many were to be found in villages which they had occupied before the Mahdiya, particularly in the northern part of the Ridge. There were a few new Arab villages, mostly along the Ridge, where overcrowding in an old village had led to some outward expansion of settlement. For the Arabs of Gedaref, and indeed for certain Western Sudanese tribes, there was still enormous prestige in the ownership of beasts, particularly cattle. They thus preferred to settle in the north from where livestock could readily move in the rainy season to graze the semi-desert grasses, away from the insects in the south. Although many Arabs were peasant farmers, with all the problems of seasonal indebtedness that this implies, others had accumulated wealth over time and placed a high value on leisure (see also Culwick 1955). Some engaged in business but many did only casual work in the dry season, using camels to transport building materials or commercial goods.

Western Sudanese predominated in the southern part of the Gedaref Ridge and formed the large majority of the pioneer settlers beside the mechanically excavated *hafirs*. Western Sudanese were particularly experienced in harvesting gum arabic in the dry season. Many had acquired rights to tap the *hashab* trees that grew wild in parts of the plains. The work was intermittent and thus could be combined with fetching water locally but, where water was scarce, some preferred to seek casual labouring jobs elsewhere.

West Africans were found throughout the area but predominated along the River Rahad, where almost two-thirds of the villages contained only West African tribes. The river, particularly in the under-utilized south, offered opportunities for fishing and for bank-side horticulture in the dry season, women being involved in drying tomatoes and peppers and processing fish for sale. Materials for crafts such as mat weaving – many West Africans were skilled crafts people – were also more available in the south. In the north, proximity to irrigation schemes on the Blue Nile offered employment for men and, indeed, whole families in the dry season. In general West Africans were not keen to spend time getting water to the detriment of income generation. Moreover, many had not tied up capital in acquiring animals that could fetch water from further afield. Thus they preferred to remain settled throughout the year beside decent water supplies or to engage in regular seasonal employment elsewhere. Tables 5.2 and 5.3 summarize the situation.

Generally relations between the diverse people that made up Gedaref's population were good. But at the Rahad there was sometimes racial tension. Typically this occurred when nomadic stock mistakenly or misguidedly used pools that West African villagers had reserved for human use. Elsewhere in Gedaref, West Africans, as foreigners, generally lived under Sudanese sheikhs, but here they

Table 5.2   Dry Season Water Supplies and Ways of Life According to Tribal Group

	% of villages practising these ways of life		
	Local Arabs	Western Sudanese	West Africans
Fully settled	54	71	70
Making local movements to obtain water in dry season	29	22	18
Transhumant	17	7	12

Source: Graham 1963: 453

Table 5.3   Dry Season Water Supplies and Ways of Life According to Environment

	% of villages practising these ways of life			
	Ridge	Hills	River	Plains
Fully settled	65	51	100	52
Making local movements to obtain water in dry season	31	20	0	42
Transhumant	4	29	0	6

Source: Graham 1963: 455

assumed village sheikhships themselves and thus, particularly in second or third generation villages, felt that they could exert their authority.

An Assessment of Ways of Life in Gedaref

Transhumance and Other Seasonal Migrations

Examples of transhumance show how different ethnic groups reacted to seasonal unavailability of water. Interestingly families were prepared to divide, perhaps for several weeks, so that women and children, whose role in agriculture was limited (except among some Western Sudanese), might have adequate water supplies. Traditional Arab transhumance consisted of movement of entire villages to an established riverine camp in the dry season. Such movements were remarkably flexible. A village could move earlier or later without undue upheaval. If water supplies became short before harvest, women and children, accompanied by a few elderly men as chaperones, could go ahead to the camp, and, if early rains failed to fill wells, return later. Animals and people were together; milk and meat were available. The move, made out of necessity, could also be turned to advantage because men had leisure to engage in dry season pursuits.

But the southern part of the Rahad was unsatisfactory in the rains, unhealthy for humans and animals, with crop yields reduced by weeds and birds. So there was a reverse migration towards the hills where good land and drier hill foot sites were attractive. (For health reasons some Arab women and children still moved north right out of the area with their livestock in the rains.) Desire for new land was also common in the Ridge. It was quite usual for groups of farmers to move out into the clays – perhaps where there was a temporary pool or even carrying water with them – to spend several weeks at a stretch living beside their fields. Only if water supplies were exceptionally good did whole families or communities go.

Dry-season transhumance was also undertaken for purely economic reasons, water availability having little to do with the decision. For example, West African pilgrims often picked cotton on the vast, irrigated Gezira Scheme in Blue Nile Province, those most involved being from the River Rahad. Sometimes men went alone, sometimes whole families went, an incentive being the grain rations as well as the pay provided.

Dry Season Local Movements to Obtain Water

Daily journeys to obtain water in the dry season enabled villages to remain settled throughout the year and were common. In most cases a certain prosperity was required either to pay for water to be fetched or to permit seasonal earnings to be limited. Moreover, although staying put brought very real benefits to households, the authorities were seldom willing to provide services to villages where all-year water was not assured.

Permanent Settlement

Women probably had a special interest in permanent settlement – it was they who were likely to benefit most from a permanent home, which could be improved as circumstances permitted. This stimulated an increasing demand for girls' education and for health facilities, which particularly benefited young children. At least in one case, when a new *hafir* was constructed, men used it for several years without bringing their families. Only when its reliability was established did some families move on a seasonal basis, but most retained their permanent houses on the outskirts of Gedaref Town.

The Mechanized Crop Production Schemes (MCPSs)

When the study was undertaken the government MCPSs had been in their existing form since 1954, when 1,000 feddan[3] holdings were rented to private individuals, with the necessary machinery and capital (Ministry of Agriculture 1954, Graham 1967). This put the MCPSs out of the reach of all but the most wealthy, who were overwhelmingly absentees from outside the area. The MCPSs required considerable labour. A holding needed about two hundred workers for two months for the harvest, and again these were mostly Westerners from outside the area, since local farmers were busy in their own fields. A very few peasant farmers had received 40 feddan holdings during an earlier share-cropping phase of the scheme. Many more local people were adversely affected. Some lost gum arabic holdings, others had their land consolidated, and the price of sorghum was pushed down, since the schemes were already producing most of the district's grain. However, some farmers with capital, mostly local Arabs or well-established Western Sudanese, began to buy or rent tractors themselves and to employ labour on a larger scale. Gedaref Town, whose population had grown from an estimated 17,000 in 1955 to an estimated average of 50,000 in 1961, boomed. Pressure on the town's wells – the population swelled to perhaps 70,000 in the dry season when agricultural labourers drifted in looking for temporary work – prompted planners to consider investment in reservoirs or a piped water supply from the Atbara river.

Pioneer Villages on the Plains

Abu Hamir typified the pioneer qualities of *hafir* villages in the clay plains outside the MCPS. A mechanically excavated *hafir* had been constructed in the dry season of 1958 to open up new land to assist farmers from overcrowded parts of the hills and settlers were rapidly attracted to it. At the end of 1960 the village consisted of about 150 houses.

Social organization was still evolving. The officially appointed sheikh was an absentee and informal leaders of the largest groups assumed de facto authority. Settlers originated from as far afield as Senegal and Eritrea; two-thirds of adult men were born in Chad or Darfur compared with 12 per cent who had been born in Gedaref District. Twenty-one tribes were represented. Half the men were unmarried – mostly youngsters who had been in the District for less than five years and were strongly motivated to earn money; for them fresh farmland compensated for rough living conditions. A quarter of marriages were mixed, tolerated here but uncommon in older settlements. Barely 10 per cent of households reached the District average of six persons.

The village was unpleasantly muddy and fly-ridden in the rainy season. Facilities were basic. There were two shops, a laundry, a butcher's stall and two guesthouses. Women from Western Sudan ran five beer parlours (four of them illegally) and engaged in prostitution as well as farming in their own right. There was no regular motor transport; the nearest dispensary and schools were five miles away; the few children who went to school had to board; and the village was cut off for weeks at a time in the rains. This unlikely village became the area's heroine and focal point in a year of water scarcity.

Drought

1961 was a year of drought. Throughout Gedaref District rainfall in 1960 had been at least 25 per cent below the mean. About half the mechanically excavated *hafirs* failed to fill properly. Some remained empty because storms were not sufficiently heavy to cause the seasonal streams that fed them to flow. Everywhere hand-dug *hafirs* dried up about a fortnight, and seasonal wells five to six weeks, earlier than usual. Moreover the first rains of 1961 were late, so that the period of water shortage extended to five or six months compared with the normal three. Ways of life had to be altered – unexpectedly, inconveniently and expensively.

Response to Unexpected Water Shortage: A Case Study

A snapshot of twelve villages in the granite hills of Qala' en Nahl illustrates the response (Graham 1973). Many mechanically excavated *hafirs* had been constructed in the area in the 1950s. Some were designed to supplement water supplies in existing villages, others to open up areas of virgin clay plain. These latter rapidly attracted immigrants and, of the twelve villages concerned, six were old and six were new. Because of the improvements in water supplies, most people in the old villages had become permanently settled. A minority still practised traditional transhumance. Large herds still went to the river, since the authorities reserved most local, new *hafirs* for domestic use. Two new, outlying agricultural communities sent their dependants to spend the dry season in neighbouring villages with more satisfactory water points. A few people fished at the river; another group went to pick cotton.

The contrast in this season of poor rainfall was acute. Two of the main *hafirs*, one at Ban and one near Bea, failed to fill. There was a marked increase in transhumance. For most this meant a resumption of a traditional migration, a return to dry season riverine campsites 24–48 kilometres away, which had been abandoned up to ten years previously. For these villages the move was familiar. But now its characteristics were different. Instead of the customary mass exodus of the whole village, there was a straggly movement over several months, families hesitating in their reluctance to leave their homes. Sometimes several families banded together to hire a truck to make the move, as the number of baggage animals was now fewer than formerly. Some families avoided transhumance by paying prolonged visits to relatives in town or in better-watered villages. Some new villages were forced to make a makeshift emergency evacuation for the first time; the absence of kin locally and the lack of a traditional dry season campsite greatly increased the hardship of the move.

There was much individual reluctance to migrate. Many people with donkeys and camels used them to fetch water from the remaining *hafir* at Abu Hamir up to eight miles away, but three communities more than ten miles away were too far off to use its water. The average distance travelled by those fetching water in this dry season was five miles, compared with two and a half in normal years. A few people, usually the very rich or very busy, were prepared to pay for water to be fetched. But while the cost of water delivered from a water point within or fairly close to a village was normally about 0.4 piastres a tin, the cost of delivering it from five to six miles away was around 3 piastres.

People without animals (the poor in the older villages and many recent immigrants), together with households without able-bodied men, were unable to fetch water. Some, including families of women and children from Ban, moved to squat in empty huts or temporary shelters on the outskirts of Abu Hamir. Others, often elderly widows, stayed up all night collecting the trickles of water that still seeped into some of the wells. Yet others dug out long-neglected *jamams*. The water yielded by these sources was so scanty and saline that consumption abruptly declined. Families were already frugal in their use of water but now many made stringent economies, using a half or a third as much as usual to the detriment of health, hygiene and living standards.

Ban, the local service and administrative centre, was particularly hard-hit. Most of the shops in the market closed, the boys' and girls' schools shut months ahead of the summer vacation, and the dispenser was unsure whether to follow the bulk of his clientele who had migrated to the river. But because the population was prosperous, and politically important enough to warrant occasional deliveries of water by government truck, many people remained settled – at heavy financial cost.

Compared with these semi-deserted villages, Abu Hamir was bustling. It provided water for the remaining populations of six other villages. About 170 donkeys and camels called each day to fetch water, along with a tractor and trailer from Ban and a donkey-drawn water cart from Bea. Around 100,000 litres were withdrawn daily and occasionally much more: for a commercial or government truck might call to fill up with water to sell to other settlements further away. Abu Hamir's own beer parlours flourished. Several shops from villages short of water re-opened in temporary quarters in Abu Hamir. Ban's prostitutes rented a hut for the remainder of the dry season. Abu Hamir's own population was augmented, not only by temporary squatters but also by twenty-five new families from the north who, attracted by the apparent abundance and reliability of its *hafir*, planned to move there permanently.

The Effects of the Drought

Economically, not merely this area but the whole of Gedaref was seriously affected. Many wells dried up before the harvest was fully gathered in. The normally profitable pursuits of gum-picking and timber extraction were largely abandoned. In marginal settlements, houses went un-repaired since either people had become transhumant or menfolk were fully occupied fetching water from neighbouring water points. At the rivers Rahad and Atbara, the pressure of livestock was unusually great. More than the usual number of animals congregated there and much earlier in the season than was customary. The riverine littorals were over-grazed; stock lost condition, some starved to death and at the Rahad hungry cattle destroyed vegetable gardens. People who had had to purchase water unexpectedly, mainly those without transport animals who were already poor, often incurred debt. The wealthy, who normally paid for their water to be fetched, were faced with increased charges with a consequent drain on their savings. The situation was exacerbated because the first proper rains of 1961 did not fall until the end of June. This affected income from livestock and crop production in the coming season.

Generally, existing social and administrative structures stood up well. But a handful of unscrupulous people exploited the situation, charging exorbitant rates (5 or even 7 piastres a tin) to fetch water for those unable to do so. Some private truck drivers, turning to water delivery, were unreliable, going wherever the profit was greatest. Gedaref's water carriers went on strike for the right to increase charges, because slow well recharge meant that they could deliver less water daily. The Town authorities eventually permitted an increase. The local councils tried to ameliorate the situation, trucking water at reasonable rates on a regular basis. For example, Gedaref South Council supplied over twenty villages in the

Ridge with water for several months. To help families avoid loan sharks, this Council provided credit against the security of women's jewellery.

Given the range of responses to the drought it was hard to assess whether, overall, women or men were most affected. Women had to cope with scarcity of domestic water (and in some cases a reduced food supply) or with squatting in uncomfortable temporary conditions, sometimes in a strange environment, if an emergency move was made. Men had the additional physical demands of fetching water from further away, while in some cases trying to maintain dry season occupations, yet needed to do whatever they could to conserve their strength for the next farming season.

The drought emphasized the known vulnerability of the area to rainfall irregularity – but ways of life were to some extent able to cope. The general availability of transport animals and the frequent lack of demanding dry season occupations enabled the people of well-established villages, who were mostly Arabs or Western Sudanese, to fetch water from neighbouring water points. These factors, together with the simplicity of building materials and the absence of exclusive grazing rights, also meant that a seasonal migration could be made with little prior planning. With a general labour shortage in Sudan's irrigation schemes, West African pilgrims could find employment elsewhere. The communities who suffered most were probably those of scattered Western Sudanese, newly settled in the clay plains, remote from other water points and generally without influence with the authorities. There was thus a certain ebb and flow of settlers and, indeed, settlements in this marginal water environment. When new *hafirs* or bores were developed, the reaction was usually rapid, the first settlers arriving with the construction gangs: though after 1961 there was greater caution. When, on the other hand, an established village's water supplies deteriorated, it was usually several years before villagers recognised that the failure was likely to be permanent and abandoned the site.

Attitudes to Water

In Gedaref, with the influence of an Islamic cultural heritage, water was generally regarded as a public good, available to those who needed it, even in times of shortage. Consequently much fuller use could be made of the scanty water resources and much better provision made for vulnerable members of the community. Two examples suffice. When a village water point showed signs of running short, the wealthier villagers often used their animals to fetch water from further off, their action allowing the poor or vulnerable, typically female-headed households or those without an able-bodied man, to use such water as seeped slowly into the well, thus enabling them to remain in the village throughout the dry season. Similarly, where a village had split as a result of a quarrel, the splinter group often lacked a water point of its own and continued, albeit shamefacedly, to obtain water from its original water source. (Of nine villages in the Gedaref Ridge that had recently divided, in only one case had the argument been over water and seven continued to use their old water sources.)

The giving of water, originally regarded as an act of religious charity, often assumed the status of a legal obligation. At the domestic level, a visitor would automatically be offered a large bowl containing perhaps a couple of litres of water, as a sign of hospitality. The accepted order of precedence was: thirsty people, the owner of the water point, travellers, local people, then animals belonging first to the owner, then to travellers, then to local people and nomads (see also Caponera 1954). As a gift of God, water for humans was free, although there might be a charge for drawing and delivering it. Only in Gedaref Town, where urban and commercial values had come to prevail, were private well-owners permitted to sell water to the public.

Water for people always took precedence over water for livestock, although a couple of milch goats and a donkey were usually included in a family's right to water. Some mechanically excavated *hafirs* were reserved only for people. Where a well was known to run short in the dry season, animals would probably never be watered at it. They drank from pools in the rains, then hand-dug *hafirs*, later perhaps at a mechanically excavated *hafir* or a bore beyond the village, then finally at the rivers Atbara and Rahad. A few wells and hand-dug *hafirs* were traditionally reserved for particular nomadic groups. Water for animals generally took precedence over water for irrigation and the few owners of irrigated gardens would usually contract to water stock for a fee. Hand-dug *hafirs* were regarded as the property of the villages concerned but mechanically excavated *hafirs*, which were free, and deep bores, where livestock were charged, were developed for the broader community, enabling some flexibility in livestock drinking patterns.

By the 1960s most water points were constructed by local or central government rather than individuals. Their siting was influenced by local politics, with the result that pressure was brought to bear to improve permanent water supplies in the areas of the densest, oldest settlements, as well as to develop adjacent water points that would open up new cropland and grazing. For example, one Arab village in the far north of the area had persuaded Gedaref North Rural Council to provide it with a new well-field and a greatly enlarged hand-dug *hafir*. But the responsibility for constructing deep bores and mechanically excavated *hafirs* lay ultimately with central government (through the Department of Land Use and Rural Water Supplies), which increasingly emphasized conservation, whether for cultivation, forestry or grazing (Robertson 1950, Bayoumi 1961). There was thus a certain tension over priorities. These authorities were male-dominated. Women took no part in public life and the chances of them influencing water supplies through their menfolk diminished as responsibility became less local. Only in relation to routine cleaning and basic maintenance of wells, where women had a special interest and which were tasks still commonly organized by village sheikhs, were they able to exert influence.

Conclusion

The research showed clearly that, at all levels, water was a social as well as a physical construct and that ethnicity was the main variable. Cultural and religious gender norms defined who might fetch water in different circumstances, and how scarce water resources would be allocated. Social perceptions of appropriate standards of living, partly at least defined by women, and affiliations with place determined the amount of water families considered necessary – and these considerations might prevail over economic advantage. Seasonal migrations reflected underlying attitudes to women and children – the desirability of women accessing water and the need to provide reasonable home conditions for children. It was striking that gender permeated so many aspects of the research: gender was in a complex relationship with the physical environment and with infrastructural and institutional changes.

General concepts of the role of government *vis-a-vis* the role of the community, as well as technical considerations, influenced who sited, who developed and who maintained water supplies. And, in circumstances where rain-harvesting techniques (*hafirs*) were so critical, maintenance (that often-forgotten aspect of technology) assumed major importance.

Despite considerable government effort to encourage permanent settlement, most villages in Gedaref remained vulnerable to the vagaries of annual rainfall. Flexibility and adaptability were assets in times of marginal water supplies – in 1961 those most accustomed to transhumance were far less inconvenienced than those attempting a more sedentary existence. The disruption caused by unexpected shortage was all the harder to cope with if services such as schools and clinics were affected, emphasizing the difficulty of achieving development and increased living standards. The physical constraints on water supplies, at a time when population was increasing and technology related to water was improving, were a salutary reminder that lack of potable water was one of the main environmental issues in Sudan and the African savannah more generally.

Notes

1. The results of the first population census of 1955 gave the population of Gedaref District as:

- All Arabs (including those from the west)   40.5 per cent
- West Africans   35.8 per cent
- Others (mainly Western Sudanese negro peoples) 23.7 per cent

2. A small allowance has been made for spillage in the case of buckets and tins and for evaporation in the case of other containers.
3. A feddan is 0.42 hectares or just over 1 acre.

References

Bayoumi, A. (1961), *Some Problems of Land Use and Rural Water Development in the Republic of the Sudan*, Khartoum: Ministry of Agriculture.

Caponera, D. (1954), *Water laws in Moslem countries*, Rome: FAO.

Culwick, G. (1955), *A Study of the Human Factor in the Gezira*, Wad Medani, Sudan: Gezira Board.

Government of Sudan (from 1905), uncatalogued archives: Khartoum, Gedaref, and Kassala.

Graham (Coles), A. (1963), 'Rural Water Supplies in Gedaref District, Sudan', Unpublished thesis, London University.

—— (1967), 'Man–Water Relations in the East Central Sudan', in M. Thomas and G. Whittington, eds, *Environment and Land Use in Africa*, London: Methuen.

—— (1973), 'Adapting to Water Shortage in a Year of Poor Rains: A Case Study from the Sudan', *Savannah*, 2: 121–5.

Mather, D. (1954), 'Aspects of Migration in the Anglo-Egyptian Sudan', Unpublished thesis, London University.

Ministry of Agriculture (1954), Working Party's Report on the Mechanical Crop Production Scheme, Khartoum: Sudan Survey Dept.

Robertson, A. (1950), *The Hafir – What – Why – Where – How*, Agriculture Bulletins, Khartoum: Ministry of Agriculture.

6

Gender Mainstreaming in the Water Sector in Nepal: A Real Commitment or a Token?

Shibesh Chandra Regmi

Introduction

Water is a finite resource and needs to be effectively managed. Women play traditional roles as water managers, store keepers, users and actors, and they, together with men, have the right to development; women should actively engage in the management of depleting water resources. However, this demands a profound commitment from those engaged in the water sector, to ensure attention to gender at all levels, especially at the organizational level, as this is where the relevant policies are formulated and practices are shaped.

This chapter is based on research, carried out over two years, that looked at rural water supplies in Nepal provided by an NGO (Nepal Water for Health – NEWAH), a bilateral project (Rural Water Supply and Sanitation Project – RWSSP), and the government (Fourth Rural Water Supply and Sanitation Sector Project – FRWSSSP). The NGO was initially funded by WaterAid then by DFID, while the bilateral and the government projects were funded by the Finnish International Development Agency (FINNIDA) and the Asian Development Bank (ADB) respectively. These agencies had implemented both gravity schemes and point sources in different parts of the country. A critical and comparative analysis of the efforts made by these agencies, in their institutional policies and practices, revealed that their gender work was ad hoc, event based and one-off and not well backed up by policies and resources – although the NGO proved more gender sensitive than the others. It is hoped that presenting the findings and analysis will help the water sector to become more gender sensitive in addressing the 'practical and strategic gender needs'[1] of rural Nepali women.

The Position of Women in Nepal

In most caste and ethnic groups of Nepal, women have a lower status and heavier workloads than men. In most rural populations, parents often give their girl children less food, less education and fewer opportunities for self-development. Various censuses have shown that the sex ratios are not in favour of women. Nepalese women do not have control over their own bodies, as indicated by the very high fertility rate of 5.6, due to society's preference for boys over girls. In Nepal, the infant and child mortality rates, under five mortality rate, crude death rate and life expectancy are all in favour of boys/men signifying neglect of female infants and children, since medical evidence suggests that girls are stronger than boys during infancy and early childhood.

The literacy rate for women is only 35 per cent. The patrilineal system of land inheritance means productive assets are usually owned and controlled by men. Despite their substantial labour contributions to the agriculture sector, less than 50 per cent of women are recorded as economically active. Women's participation in politics, bureaucracy, judiciary, and other constitutional bodies is also extremely low (Acharya 1997, Regmi 1989 and 2000, Singh 1995, UNICEF 1997).

It is in this context that development work is undertaken in Nepal. It is likely that any development agencies, internal or external, will reinforce the existing gender and power relations in their work if they do not make special efforts to address the gender inequalities prevalent in Nepalese society.

Approaches and Perspectives

The research focused on gender mainstreaming, which is internationally recognized and involves the integration of gender issues into all policies, planning and practices of an organization. It involves enabling equal participation in development, and equality of access to and control over institutional resources. Gender mainstreaming demands that the existing male bias in policies and practices be addressed, and men and women be treated equally at work. The research drew on several frameworks and gender models to develop a methodology for assessing how far gender mainstreaming had been achieved, and this was applied systematically across all three agencies (Macdonald et al. 1997, Swieringa and Wierdsma 1992, Moser 1993, Goetz 1997a and 1997b). The findings from the selected agencies, and the implications of their different approaches to gender, are presented under headings reflecting the key criteria developed for measuring organizational progress. These are the institutional policies (policy formulation, vision, mission and objectives, and personnel policies); organizational structure; organizational culture; gender training; resource commitment; project implementation; and external relationships.

Although some attempts have been made to address gender mainstreaming in the water sector in Nepal, it has been little understood and applied only on an ad hoc basis; it is seen as something new, strange, and even unwanted in the water sector. There has been no serious examination to see where the process of gender mainstreaming has gone wrong and what lessons could be learnt for the future; much remains at the level of rhetoric.

Key Findings

Policy-making within the Organizations

The process of policy formulation critically affects policy outcome; the more participatory the process, the better the results of the policy. Experience (Hadjipateras 1996, Moser 1993, Hamerschlag and Reerink 1998) shows that the policies formulated with the involvement of women and men, through a democratic process, result in better outcomes at all levels.

Though the NGO staff said they felt positive about the need for gender balance in policy-making bodies in their discussions, their strategic plan 1998–2002 was silent on this matter (NEWAH 1998). In the government project, the only two women engineers (out of thousands of male technicians) were not included in policy-making. Similarly, in the bilateral project, the only woman present in the policy-making body failed to influence the decision-making process. Women in junior positions were always bypassed in meetings in all agencies; even when they were called to attend they were not consulted. As a majority of the junior women had only basic education they could not even read the policy decisions, which were written in English.

Aims and Objectives

Gender is a political issue; it is about balancing the power between women and men. The vision, mission, aims and objectives of an institution reveal the power given to women and men and the distribution of organizational benefits. Only by having a clear understanding of women's different roles and needs, and their relative disadvantage in relation to access to and control over resources, can organizations develop gender sensitive objectives that entail increased equality and empowerment for them.

In all three organizations the aims were set out within a broad commitment to supporting 'community', 'users' and 'beneficiaries', terms that in fact hide the inequalities between and different needs of women and men. These 'gender neutral' terms overlook certain realities: for example, only one to two women in the bilateral and government projects were included in the water user committees of between eleven and thirteen members, because the focus was on community or user involvement, not on addressing inequalities within the community. Benefits such as training, observation tours, paid jobs, and involvement in meetings were mostly enjoyed by men, because of the lack of commitment to addressing women's subordination within communities (Regmi and Fawcett 1999, Regmi 2000). Even though the women in the government project in the Hile area secured access to safe water near their homes, their workloads increased significantly because the men expected more water to be available at home but had stopped fetching it now that it was conveniently located. Due to the inappropriate location of tap stands women were obliged to carry water to their homes to perform their private ablutions, whilst previously they had privacy at the old water sources.

There was a lack of articulation of aims such as reducing women's labour, increasing their income, and providing opportunities to participate in various training programmes in each agency. As a result, work to achieve these changes for women was not prioritized, and their needs were subsumed under the 'community approach'.

Personnel Policies

From a gender perspective two important issues – recruitment policy and working conditions for staff – need to be analysed in any institution to ensure it can attract and retain women employees.

In the NGO and bilateral agency one extra statement 'women are encouraged to apply' was added to advertisements for staff; nothing else was written in the vacancy announcements to attract women candidates. The agencies failed to receive enough applications from women candidates, e.g. the bilateral project received only two applications from women as against thirty-eight from men when it advertised the post of District Support Advisor.

The reasons for this are complex, and go beyond the wording of advertisments. Due to their domestic responsibilities girls, in general, start and finish their school and college education later than boys. Even when they start at the same age they often finish later because they frequently miss school to help their parents, especially mothers, at home. Subsequently, women can find themselves too old to apply for vacancies. In addition far fewer girls than boys start or complete their education, so the pool of females to recruit from is far smaller.

In all agencies the interview panels were made up of mostly men. More men were recruited because the interview panels often perceived the biological responsibilities (child care for example) of women to be their personal problem and an impediment to performing well at work. Panels also had a tendency to think that men were more appropriate for technical jobs in the water sector. The high number of male staff in all institutions, especially in the engineering sections, confirmed this argument.

Hamerschlag and Reerink (1998) note that family-friendly work policies enable workers – both women and men – to balance their work and family responsibilities more easily. Since women are often the primary caregivers in families, providing family-friendly work options often has a greater impact on women, enabling them to take on more senior-level positions in an organization without negatively affecting their ability to care for their families. All the institutions had some facilities for their employees, but these were highly variable. The NGO offered paternity leave of seven days, something not yet provided by many organizations in Nepal. In contrast, the bilateral project did not have maternity leave or funeral leave for any employees, even though these are the facilities provided by the majority of organizations in Nepal. The reason given by management was that staff were paid well.

In the government department, staff were given facilities in accordance with the Nizamati Sewa Niyamawali (Civil Service Regulations), which appear to be more gender sensitive than the other two organizations (HMG/Nepal 1998). Women are entitled to maternity leave of sixty days twice in their service; the men are given funeral leave of fifteen days per annum. If both husband and wife are in government jobs they are placed in the same area if possible. Female staff are exempted from the age barrier for posts if they work in government for some time, and they have only six months probation against one year for men; female employees are considered eligible for promotion before one year of service, the minimum criterion for men.

However, the lack of any extra facilities for childcare for women while under-taking fieldwork (in all the agencies), the refusal to allow women on maternity leave to use accumulated leave from the previous year (in the NGO), the lack of maternity leave for women employees and crematory leave for male employees (in the bilateral project), and the same age of retirement for women and men employees though their start date could be different for various reasons (in the government) all indicate improvements could be made. Even when these agencies attract women to join them, they have difficulty in retaining them (Regmi 2000). In agencies which try to be flexible around working hours for breastfeeding, or time off for fulfilling household and community responsibilities, or allowing women not to go to the field during pregnancy, these conditions are often not enshrined in policy documents and are applied informally, so many women may not enjoy them.

Results of Discretionary Power

One junior female staff had to return to the office within twenty-two days of her delivery in the bilateral project, which is rare in the Nepalese context. As a result, her baby was compelled to rely on bottle-feeding by others. Some others in the NGO and the government had to resign as neither their maternity leave was adequate for them nor were they allowed to take extra leave. But when there was only one senior woman in the NGO she could enjoy flexible working hours. Also in the NGO, one woman, being the section head, was receiving transportation facilities but another woman, though working at the same level, was not because she was not the section head.

Staffing Structure

The higher the level of female representation in an organization, the greater staff's commitment to integrate gender into institutional activities. The experiences of ACORD programmes in Gao (Mali) and Gulu (Uganda), and of Novib, Hadjipateras (1998) and Macdonald et al. (1997) show that one guarantee of gender sensitivity in an organization is the presence of a significant number of women, among whom are strong, gender sensitive women, committed to women's empowerment and gender equality. A critical mass is needed, about 30–35 per cent female representation according to the UN.

The proportion of women staff was around 17 per cent in the NGO, 8 per cent in the bilateral project and 14 per cent in the government (including regional and central office staff and all permanent and temporary staff at DWSO). Though the number of women was gradually increasing in all organizations, none of them had been able to break the traditional gender division of labour where men worked mostly on the 'hardware' side and women on the 'software' side. There was only one woman technician out of twenty-five technicians in the NGO, no female technicians out of eight in the bilateral project, including those in the counterpart organization, and five women engineers and six overseers out of thousands of male technicians in the government project. This gender imbalance contributed to a failure to formulate gender sensitive policies, retain women staff and ensure women's presence during project implementation. Women's needs were overlooked much of the time (Regmi 2000, Regmi and Fawcett 2002).

The NGO had made some ad hoc decisions to change the traditional gender division of labour by recruiting some men (instead of women) to the position of health supervisor, and one woman to a technical job (instead of a man). However, the fear was that unless enough women were found to compete for technical positions, the practice of recruiting men to the health sector might even close the door on these limited opportunities for women.

Alongside the negative attitudes of the senior management, which blocked the recruitment of women to technical positions, the unavailability of technically competent women was another constraint. The number of girl students applying for engineering studies, passing the entrance test and actually enrolling was lower than the number of boys in both 1997 and 1998 (Table 6.1). Similarly, though the absolute number of girls receiving admission on the basis of their performance increased in the second year, the lower proportion receiving admission based on merit indicates they were having difficulty competing with boys. This was because of their relatively poor educational backgrounds caused by parental bias in providing good education to sons over daughters. The trend is the same in other engineering colleges across the country.

Table 6.1 Numbers of Boy and Girl Students Applying for Civil Engineering Courses at the Nepal Engineering Campus in 1997 and 1998

Number of students	1997/98			1998/99		
	Boys	Girls	Total	Boys	Girls	Total
Applicants	856	108	964	1261	203	1464
Entrance test passers	282	27	309	875	154	1029
Actual enrollers	92	8	100	131	13	144

Source: Nepal Engineering Campus, Tribhuwan University, Kathmandu, 1998

These findings suggest that the institutions working in the water sector, where women's role at the community level is central, need to specify in their policy documents that a certain number of their staff will be women. To achieve this they will need to provide scholarships to girls, provide on-the-job training, and good conditions once women are recruited.

Organizational Culture

Organizational culture is the personality of an organization; if the organization's structure is the body, its personality or soul is the way people deal with each other and the values and beliefs that are dominant. Organizational culture determines the conventions and unwritten rules of the organization, its norms of co-operation and conflict, and its channels for exerting influence (Macdonald et al. 1997). It also shapes the distribution of power and authority between male and female staff that in turn shapes their behaviour.

In the NGO, staff were observed to be friendly, eating together, teasing each other, and maintaining informal relationships. In the bilateral project the management, consisting of mainly male foreigners, had close relationships with male senior Nepali staff; there was a big gap between the management and both the junior staff and the only senior woman in the project. The relationships between the senior and the junior staff were formal and hierarchical in the government project, where position in the hierarchy dictated staff behaviour.

In all agencies, women staff felt less powerful than their equivalent male counterparts. They had the impression that because they had to be absent more than men to fulfil their reproductive roles, and because they had negligible representation at the policy-making level, the male-dominated management listened more to the men. The lack of provision for separate lavatories and childcare and transportation facilities for women staff regardless of seniority;

women's involvement mainly in the software side; the lower value given to women's needs and perspectives; and the limited opportunities that women – particularly those in government as opposed to those in NGOs – had to express their opinions clearly implied a lack of understanding of the gender differentials. The unwillingness of agencies to address those differentials in their structure, administration and physical arrangements strongly indicated a women-unfriendly organizational culture.

The majority of staff in all agencies, though especially in the government project, had a weak understanding of gender issues in water. For many, gender meant women's issues and women's involvement in water supplies was felt to be important mainly because they are the primary users. They saw the need for women's participation during the implementation phase (providing labour), during the post-construction phase (cleaning and protecting the water sources), and in maintaining hygiene at home. For most there was no gender issue, no conflict of interest or difference between women and men in accessing and controlling water supplies and benefits. Out of the many staff interviewed in these agencies only three people (two women and one man) in the NGO, and two people each in the bilateral and government projects (one woman and one man in each) saw women's involvement in water as a means to their empowerment.

However, a majority of the staff at the NGO were in favour of change, recognising the need for greater gender equality and that gender sensitivity was essential to achieving this. NEWAH introduced of a number of new approaches following the research, well described in Chapter 11. These included workshops and gender training; strategy review and new project implementation procedures; new personnel policies; frequent meetings and field visits; employment of more women; and the introduction of pilot gender and poverty (GAP) projects. Despite some fear among the men that changes might affect them in terms of their power, they accepted that organizational culture was a continuous, dynamic and fluid process, involving interactions and renegotiations. Analysing the gender challenges in communities and internally convinced staff that these changes were inevitable and necessary. In contrast, the bilateral and government projects continued to follow top-down blueprint approaches.

Another determinant of a gender sensitive organizational culture is the space given to change agents to bring about gender reforms. According to Macdonald et al. (1997) such change agents need three things: modest aims and ambitions since changing the personality – the soul – of an organization can be a difficult and painful process; an understanding of the organizational culture to help find clues about how it can be changed; and flexibility in strategy to determine what works best. In the NGO, it was the male Director, and the three senior women staff working in the health section who were the key gender change agents. In the bilateral project, it was the male institutional advisor and the female health advisor who introduced gender activities in the organization; these change agents were relatively senior. They were diplomatic and pragmatic, building alliances and strategies both with the staff and the management; they negotiated gender policies, not just for the benefit of one group but for the overall organization. By contrast, no change agents were observed in the government project, though many young officers at the bottom of the hierarchy were aware of the need for gender sensitivity in the water sector. However, their low status meant they lacked influence in policy matters and they had no forum in which to even raise gender issues.

Gender Training

Gender training is important for raising awareness and to provide concepts and techniques for programme workers; however, the long-term impact of training is easily reduced if it is provided only to a few staff, is one-off, or if resources for follow-up activities are lacking. From the experiences of ACORD and Oxfam, Hadjipateras (1997) and May (1997) it is clear that gender training, if organized in a timely manner and undertaken regularly, can bring positive changes in staff attitudes and lead to better development projects.

NEWAH organized a five-day-long gender workshop (the first in its history) for senior staff – six women and twelve men – in January 1999, and a week-long training of trainers for its thirty-one staff – seven women and twenty-four men – in June 1999. The training continues and has been instrumental in integrating gender into the health and sanitation guidelines, the design of latrine and water posts, and the recruitment of daily wage staff. The bilateral and government projects had no formal training on gender, and no plans to introduce it, even though senior staff said it was necessary.

Resource Commitment

The lack of adequate resources, both capital and human, has a tremendous influence in realising any project objectives. Jahan (1995) and Hadjipateras (1997) note that the lack of resources often leads to agencies' failure to promote gender work.

The NGO had substantial funds available for gender activities from various sources including WaterAid (UK and Nepal), DFID, other donors and its own core funds. Yet it had not allocated any funds for gender activity in its 1999 annual plan and only £3,200 (0.5 per cent of its total budget of approximately £0.6m sterling) was spent on gender activities that year. The bilateral project did not have any funds allocated for gender in Phases I and II though a budget was allocated for one specific activity in Phase III. The government project lacked a budget for gender-related activities in its five-year project. The lack of resources made available by these agencies shows that gender was not their priority.

The human resources allocated to gender were also sparse. The NGO had one expatriate female consultant to initiate gender-focused activities in the organization. This consultant and the author initiated the gender and poverty project in NEWAH, and a gender coordinator was assigned to work alongside the expatriate for continuity. Despite these efforts, gender did not become a priority for the NGO until later, as neither the expatriate nor the gender coordinator were members of the senior management and thus lacked influence. The bilateral and the government projects did not assign anyone to look after gender; consequently, even where the project documents demanded two women be in a local users committee, water supplies were often implemented with only one or no women in the Water Users' Committees (e.g. Hile, Belbhariya (Gajedi), and Gajedi project areas), with negative consequences. The intensity of the gender work is often determined by the presence of a designated gender person; however, if such people are not in decision-making positions the work cannot take off at full speed.

Project Implementation

All agencies claimed that they were adopting participatory approaches while implementing their water schemes. They discussed projects with local people before implementing them and the communities in turn provided labour during construction and formed the local committees to look after the implementation as well as the operational and the maintenance phases.

In all agencies one of the requirements to get the water was that the concerned communities had to file a request for it with signatures from the potential users. However, there was no involvement of women at all at this stage – in every project all signatories were men. Yet the applications were processed anyway by the agencies, instead of being returned for women's signatures to ensure women's participation from the start. Similarly, a number of criteria had been developed by these agencies for selecting local partners to implement water projects, but criteria such as gender balance of staffing, experience in gender work and the presence of people who have received gender training were missing. This resulted in male staff and male local leaders making all decisions in project matters. One added bias in the government was the attitude of the senior officials who, despite recruiting women to the post of technician because of contemporary pressure (van Wijk-Sijbesma 1985 and 1998, Bilqis et al. 1991, Baden 1993, Fong et al. 1996), were not allowing them to do technical field work so undermining their interest and competence.

Feasibility studies and detailed surveys are other important steps in the implementation of water projects. The type of information collected, who collects it, the time spent, and the analysis of gender roles and division of labour are all critical for ensuring that poor and marginalized women, together with men, are heard and their concerns met in the project design and planning. In the

NGO, where at least one woman was included in the feasibility team, meetings were called in places and at times that were appropriate for local poor women. However, the presence of women in such teams was not compulsory. In the bilateral and the government projects, no attempt was made to include women in such teams nor was gender analysis required; even in the policy documents, the focus was on technical not social or gender information. This had an impact on implementation; for example in one project the water tariff collected initially by women had to be stopped. They could not continue paying it because they had no control over household financial resources. This directly affected the operation and maintenance of the project, resulting in poor maintenance and an increase in the number of malfunctioning tube wells and tap stands. This could have been predicted had the gender analysis been carried out during the feasibility phase.

The agencies spent about a week on feasibility surveys. However, given the area covered, the dispersed settlement pattern, and the cultural factors that prevented women from leaving their homes easily, this time was inadequate for exploring the gender issues which were crucial to the effectiveness of water supplies. During the surveys, project staff informed the communities of the criteria to be met by them in order to receive water supplies, but no time was spent asking whether the poor, especially women, were in a position to meet all these criteria. All institutions made use of predetermined technologies and the women were not consulted either about the design of the tap stands/tube wells or the platforms. Subsequently, the women in some of the NGO and the bilateral projects complained that they had to bend hard while pumping the tube wells which gave them discomfort in their waist and back; they found the handle of the pump too long for their height.

External Relationships

Gender mainstreaming in an organization also depends upon its external relationships and contacts. For example, the decision to hire a gender consultant at the NGO was initiated by WaterAid; the efforts to recruit women technicians started following comments from a team of expatriates sent by WaterAid, UK who came to evaluate NEWAH in 1997. In the bilateral project there was also external influence as the management was completely foreign controlled. The decision to carry out a gender study and increase the number of women staff in the bilateral project was suggested by the missions that came occasionally from FINNIDA Headquarters to evaluate Phases I and II. The fact that the project documents of Phase III were more gender sensitive than the other two phases was the result of these external suggestions, supplemented by the recommendations of a one-off gender study carried out by an expatriate. The situation was no different in the government project, where most of the gender activities, such as recruiting sociologists – especially women – to work with community women, recruiting women workers, and hiring women water supply and sanitation technicians (WSST), were introduced upon recommendations from ADB.

Influence of External Donors in Gender Integration

A mission from the ADB/Manila recommended to the government of Nepal based on the results of the previous project that posts of women workers at district water supply offices be created by rationalizing positions within the organization and within existing budgets. The mission also advised that Department of Water Supply and Sewerage (DWSS) should improve the balance between women workers and engineering staff. The government confirmed that it saw the inclusion of women workers in DWSS as important and would ensure that DWSS complied with the Bank's recommendation as follows: i) complete the recruitment of seventy-five women workers to fill in currently vacant positions by 31 July 1996, ii) draw up a programme to increase the number of women workers in DWSOs at a scale of two in every DWSO with one engineer and four in every DWSO with three or more engineers, iii) ensure that of the pool of forty technicians to be maintained at the regional level effective October 1996, at least 25 per cent would be women workers, and iv) draw up a programme of training of women workers commensurate with their recruitment (FRWSSSP 1996).

Since all three agencies were receiving funding support from external donors, the latter were able to influence the formers' policy formulation process; unfortunately their pressure was not consistent and gender was often not a priority for them. In the NGO, where there was less bureaucracy, more flexibility, high commitment and more accountability towards the poor and marginalized, the externally stimulated gender-related changes took root more effectively than in the other two agencies.

The Impact of Projects on Gender Relations at Community Level

These discussions show that a number of attempts were made by the selected agencies to address gender issues at all levels – community, project and organization. However, the agencies' intentions were not backed up with firm commitments on policies, terms and conditions, training, resources and recruitment. The lack of serious commitment and rather ad hoc approach had implications for women both in the organizations and in different communities.

Women's participation in meetings and decision-making was low in all project communities, though less so in the NGO projects; this often led to the wrong siting of tube wells and tap stands. So despite having improved water facilities nearby homes, local women were not necessarily benefiting – some sources were located along roadsides and women were facing hygiene difficulties because of the shame of being seen by passers-by, resulting in them waiting till dark to perform their private activities. Since the government project was at a high altitude and thus cold, the women there had to take water to their homes for personal use. The undergarments used during menstruation require lots of water, so the women had to carry water many times causing unnecessary physical exertion, mental stress and heavy loss of time. The women from the NGO and the bilateral projects had to perform such activities in the evening and at night putting themselves at various risks.

The projects had made limited attempts to increase community women's access to and control over resources. The poor and marginalized women were still engaged in traditional gender activities that did not yield any substantial income and their bargaining power had not increased much. Though the projects claimed that women could engage in various productive activities due to the time saved from water hauling, this had not come about because women's workloads had not much reduced. This was partly because of the poor location of many tap stands and tube wells caused by the lack of consultation during the selection phase. Even where women saved some time, they did not undertake successful income-generating work because the communities were remote, lacked resources, and women had no prior experience of such work and lacked confidence as well as access to markets. The projects did little to address these barriers. The agencies were making only a small impact on the financial well-being of the community women through the water supplies, even though finding ways to increase their income could have had major multiplier effects, such as investment in more nutritious food for the family, better education for the children, and overall betterment of the family. The projects failed to really promote women's deep involvement in project activities, yet this could have led to higher confidence on their part and resulted in real development within communities.

The projects' impact in improving the lives of poor people, especially women, was limited. Women were still considered primarily responsible for fetching water and were spending a substantial amount of time doing it. Since they were not involved in all the project stages and felt that men were benefiting more from the projects, women, despite the fact that they were considered the primary actors in the water sector and understood the technical requirements for keeping the water flowing, were observed to be lacking in enthusiasm for the operation and maintenance of the projects. Consequently, increasing numbers of tube wells and the tap stands ran into operational and maintenance problems, causing women once again to spend more time collecting water. This affected their engagement in other productive work on the farm and in the household, and their health. The situation of poor women was particularly acute; they were rarely involved in the planning and design of the water supplies and faced real problems paying for water. They were the first ones pushed back to the traditional unhygienic water sources far away. The distribution of benefits was not equal in the project communities due to the lack of agencies' gender sensitivity.

Finally, the selected agencies were rarely successful in improving the status, confidence and profile of the community women through water schemes, though all were strategic objectives. Women continued in secondary and stereotyped activities in all project communities; only one or two were included in the committees composed of eleven to thirteen members, only men were recruited to paid positions and women were expected to fill voluntary positions, so increasing the pressure on their already overburdened lives. Training and observation tours were reserved mainly for men. Even where women were recruited for project work they were paid less than men. In projects men were involved in technical work, while women were engaged in non-technical, low-profile and traditional work.

Only men signed agreements with the agencies. The traditional gender division of labour was reinforced because all the skilled labourers hired for water supply, latrine construction and maintenance were men, and the monitoring of progress and handling of all financial matters were their responsibility. Due to inadequate time spent by the projects in preparing women, the community women had difficulty in meeting various requirements such as filling out monitoring forms, preparing reports and sending them to the funding agencies, and visiting agencies about problems they encountered in the project, causing them to rely heavily on men for everything. They were less active in the decision-making process. Their income was only nominally raised and thus there was no change in their bargaining power at home, and some even struggled to pay user fees and were forced back to poor water sources. The women lacked the confidence to visit other development agencies to initiate new work in their communities because the projects had done little to promote their involvement in wider development activities.

Conclusion: Implications for Water Organizations

Though one key aim of the gender and development approach is to challenge stereotypical gender roles, the research showed that the agencies had made few attempts at this. As a result, women staff remained second-class citizens, whose existence and identity were dependent on men – they were always identified as someone's wife/sister/daughter/mother. Their voices were barely heard and their needs and perspectives did not shape policies or practice, with one or two exceptions. They lacked resources and access to senior jobs. NEWAH's subsequent attempts to tackle these issues showed that it was more concerned than previously to make its work gender-sensitive. Its staff showed more openness and positive attitudes; the director was supportive and acted as a change agent; and there were more women staff, including a few with designated gender responsibility, and more gender training and workshops. NEWAH proved responsive to the findings of this research.

However, a critical analysis of the gender mainstreaming efforts of all agencies showed that most gender work was ad hoc, not backed up by policies, and lacking in resources and strong commitment. A number of reasons were identified by the research to explain the slow progress made in addressing gender concerns in the organizations. These were:

- A lack of consistent gender sensitivity in the donors' (WaterAid, FINNIDA and ADB) and the Nepali government's policies, objectives and strategies. The selected projects relied heavily on the policy documents of their donors and the government while developing their own strategies, and gender was not a clear priority in these.
- The strong patriarchal norms in Nepal, which lead staff to automatically re-pudiate gender-sensitive ideas and policies because of fear amongst male bureaucrats and policy-makers that they may not retain patriarchal control over the whole system. The belief that women should always be subordinated to men so as to put them in a low profile was strongly observed in all agencies.
- Strong male domination led to a lack of gender-sensitive policy formulation; a lack of recognition of women's roles and contributions; a lack of seriousness and commitment in increasing women's roles in policy-making; and the misconception that water is a technical sector and therefore men are more fit than women to work there. It is not inevitable that gender sensitive policies will emerge when women are involved in generating policy, nor is it guaranteed that good policy will always lead to the betterment of women. Much depends on how the policies are translated into action. However, the experiences of the authors already cited and those of Regmi (2000) and Regmi and Fawcett (2001) show that women's participation in policy-formulation can lead to the formulation of more gender-sensitive policies, objectives and strategies. They are an important first step.
- The agencies showed a lack of ability to understand and judge the impact of work performed by gender-sensitive staff. They also lacked understanding of the different needs and concerns of women and men in the project communities and the fact that they cannot be treated with one single approach. The belief continued that men can easily represent women and their needs in spite of the obvious problems emerging in the projects due to misunderstanding or overlooking women. Their behaviour clearly indicated their willingness to maintain the tradition of male domination.

This research shows the critical importance of developing a better understanding of the structure of the societies where water agencies have to operate. The organizational environment has historically favoured of men and the drinking water sector in Nepal is no exception. It is only recently that women in Nepal have started to work in offices outside their homes. As a result, neither the women themselves nor men in senior positions have been able to think of ways to improve the organizational culture in their agencies. Moreover, the roles that have been expected of women and men in the Nepalese societies have been culturally determined for centuries. A change in these roles demands changes in the attitudes of both women and men and requires gender sensitization work with both sexes at all levels. Until this has been achieved some positive discrimination in favour of women will help to recruit and retain women employees.

In the patriarchal society of Nepal, institutions involved in the drinking water sector need to introduce many activities and make a lot of effort to help men understand the contribution that women make in this sector; this does not happen overnight. Given the fact that Nepalese girls, especially those who come from rural areas, have less opportunity than boys to go to schools and colleges, and even when they do they receive a relatively poor quality education in public schools, it is unreasonable to expect that soon there will be equal numbers of qualified women in the market to compete with men. Equally important is to implement activities that can address the traditional gender division of labour. It has become increasingly clear that social exclusion of women can lead to poorer interventions that further reinforce structural inequalities against them and this must be tackled.

Finally, it is clear from the above discussion that the ways in which institutions work, from policy formulation to visions and objectives, from personnel policies to gender training, from the organizational structure to its culture, and the way resources are committed and projects implemented, are key to the integration of gender. The greater the gender sensitivity in these areas, the quicker the organization can mainstream gender. The issues are complementary and none of them alone will succeed in making an institution gender sensitive. For example, there can be provision for gender training for the employees but, if there is no change in the organizational structure and culture, making them more conducive to women, the training alone may not be able to bring about any major change. There is an urgent need to look at all the institutional activities in totality when trying to engender an agency to achieve a programme promoting equality between women and men.

This does not, however, mean that if an agency cannot address all these together it should not address any. A superficial understanding of these institutional issues may lead one to say that they are mutually exclusive and, hence, can be treated separately. The message is rather that the impact will be greater when all issues are properly addressed. Only a careful consideration of the issues reveals the importance of tackling them all in order to mainstream gender in an organization, which in turn can yield meaningful and sustained results in its development work on the part of the poor and marginalized, especially women.

Note

1. Practical gender needs are those women identify within their socially accepted roles in society. They do not challenge women's subordinate position in society, although they arise out of it. They are practical, a response to an immediate necessity, and often are concerned with improving living conditions, e.g. water provision, health care, employment. Though these needs when fulfilled benefit all family members, women generally presume they are their specific needs because the responsibility for them has traditionally fallen to them (Moser 1989 and 1993). Strategic gender needs are those women identify because of their subordinate position to men in the society. They vary according to context; when met they allow women to address the power imbalances in their relationships with men, e.g. the abolition of the sexual division of labour, freedom of choice over childbearing, removal of institutionalized forms of discrimination, legal rights, and equal wages. Meeting strategic gender needs helps women to achieve greater equality (Moser 1989 and 1993).

References

Acharya, Meena (1997), *Gender Equality and Empowerment of Women*, Status report submitted to the United Nations Population fund (UNFPA), Kathmandu, Nepal.

ADB (Asian Development Bank) (1998), The Bank's Policy on Water Working Paper, August, Manila, the Philippines.

Baden, S. (1993), 'Practical Strategies for Involving Women as well as Men in Water and Sanitation Activities', Briefings on Development and Gender (BRIDGE) Report No. 11, Report prepared for Gender Office, Swedish International Development Office.

Bilqis, A. et al (1991), 'Maintaining Village Water Pumps by Women Volunteers in Bangladesh', *Health Policy and Planning*, 6 (2).

Fong, M., W. Wakeman and A. Bhusan (1996), 'Toolkit on Gender in Water and Sanitation', *Gender Toolkit Series No. 2*, Washington DC: The World Bank.

FRWSSSP (Fourth Rural Water Supply and Sanitation Sector Project) (1996), ADB Project Document, Ministry of Housing and Physical Planning, Department of Water Supply and Sewerage, Kathmandu.

Goetz, Anne Marie (1997a), 'Introduction: Getting Institutions Right for Women in Development' in Anne Marie Goetz, ed., *Getting Institutions Right for Women in Development*, London: Zed Books.

—— (1997b), 'Managing Organizational Change: The "Gendered" Organization of Space and Time', *Gender and Development*, 5(1), Oxfam.

Hadjipateras, Angela (1996), 'Community Participation in Gender Impact Assessment: Lessons from recent ACORD research', ACORD (RAPP), London. Unpublished.

—— (1997), 'Implementing a Gender Policy in ACORD: Strategies, Constraints and Challenges', *Gender and Development*, 5(1).

—— (1998), 'Putting Gender Policy into Practice: Lessons from ACORD', *BRIDGE, Briefing on Development and Gender*, Issue 5.

Hamerschlag, Kari and Annemarie Reerink (1998), Best Practices: For Gender Integration in Organizations and Programmes from the InterAction Community Findings from a Survey of Member Agencies. Washington DC, InterAction, American Council for Voluntary International Action, Commission on the Advancement of Women.

HMG/Nepal (His Majesty's Government of Nepal) (1998), *Nepal Gazette*, Part 48 (Asar 29, 2055, Extra Issue 26), Kathmandu.

Jahan, Rounag (1995), *The Elusive Agenda: Mainstreaming Women in Development*, Bangladesh and United States: University Press Limited and Zed Books.

Macdonald, Mandy, Ellen Sprenger and Ireen Dubel (1997), *Gender and Organizational Change Bridging the Gap between Policy and Practice*, The Netherlands: Royal Tropical Institute.

May, Nicky (1997), *Challenging Assumptions Gender Issues in Urban Regeneration*, UK: Joseph Rowntree Foundation.

Moser, Caroline (1989), 'Gender Planning in the Third World: Meeting Practical and Strategic Gender Needs', *World Development*, 17(11):1799–1825.

—— (1993), *Gender Planning and Development: Theory Practice and Training*, London: Routledge.

NEWAH (Nepal Water for Health) (1998), Strategic Plan 1998–2002, Kathmandu.

Regmi, Shibesh (1989), 'Female Participation in Forest Resource Management: A Case Study of a Women's Forest Committee in a Nepalese Village', Dissertation submitted to the Ateneo de Manila University of the Philippines, for the partial fulfilment of Master of Science in Social Development, Quezon City, Manila, the Philippines.

—— (2000), 'Gender Issues in the Management of Water Projects in Nepal', PhD thesis submitted to the Institute of Irrigation and Development Studies, Department of Civil and Environmental Engineering, University of Southampton, UK.

Regmi, Shibesh and Ben Fawcett (1999), 'Integrating Gender Needs into Drinking Water Projects in Nepal', *Gender and Development*, 7(3).

—— (2001), 'Men's Roles, Gender Relations, and Sustainability in Water Supplies: Some Lessons from Nepal', in Caroline Sweetman, ed., *Men's Involvement in Gender and Development Policy and Practice*, Oxford: Oxfam Working Papers.

—— (2002), 'Gender Implications of the Move from Supply-Driven to Demand-Driven Approaches in the Drinking Water Sector: A Developing Country Perspective', Paper presented at the First South Asia Water Forum, organised by Global Water Partnership and Water Energy Commission, Kathmandu.

RWSSP (Rural Water Supply and Sanitation Project) (1996a), Rural Water Supply and Sanitation Project in the Lumbini Zone Project Document Phase II 1996–1999, The Kingdom of Nepal and The Republic of Finland, FINNIDA.

—— (1996b), Rural Water Supply and Sanitation Project Lumbini Zone, Nepal Completion Report of the First Phase January 1990–July 1996, Nepal, HMG/MHPP/DWSS and The Government of Finland, Ministry of Foreign Affairs, Department for International Development Co-operation.

Singh, Sabitri (1995), Statistical Profile on Women of Nepal Stri Shakti, Kathmandu, Nepal.

Swieringa, Joop and Andre Wierdsma, (1992), *Becoming a Learning Organization: Beyond the Learning Curve*, Reading, MA: Addison Wesley.

UNICEF (1997) Statistics of South Asian Children and Women. United Nation Children's Fund Regional Office for South Asia, Kathmandu, Nepal.

Van Wijk-Sijbesma, C. (1985), 'Participation of Women in Water Supply and Sanitation: Roles and Realities', The Hague, International Reference Centre (IRC) for Community Water Supply and Sanitation Technical Paper, No. 22.

—— (1998), 'Gender in Water Resources Management, Water Supply and Sanitation Roles and Realities Revisited', The Hague, International Water and Sanitation Centre, Technical Paper Series 33-E.

7

The Challenge to International NGOs of Incorporating Gender

Tina Wallace and *Pauline Wilson*[1]

Introduction

This chapter explores the different tensions and pressures that NGOs based in the UK face when trying to meet clear water targets and concrete deliverables in contexts of extreme social and economic need on the ground. Such tensions become acute as competition for funding in the UK intensifies. Demonstrating impact becomes a priority, while experience shows that working with local communities to support and sustain infrastructure and change behaviour over the long term takes time. In gender policies, NGO strategies, and much of the literature on delivering 'water for all' the disjuncture between expanding the coverage of water and engaging the community is glossed over. Yet, in reality, the rhythm and requirements of the technology and geology of water are often at odds with the pace and demands of community engagement and social change (Davies and Garvey 1993).

It is well known and widely accepted that women are the principal users of water, yet unequal gender relations, at both institutional and community levels, curtail their access to key resources and involvement in decision-making in most contexts. 'The trend in much of the Third World has been that where technologies have been introduced they have been male dominated. In effect a male monopoly of technology has developed which has excluded women (even though) the success of any rural water project rests on the acceptance and understanding of the technology by the people who ... use it' (Mathew 1991: 190). To date it has proved difficult for agencies to overcome these internal and external inequalities to ensure women fully participate and benefit. This is even though the commitment of the water sector to include women, to promote women's empowerment, and to ensure the access of all to water has grown.

The challenges for agencies working in the water sector are many and complex, including organizational as well as programme changes, technical as well as social and economic issues. In relation to issues around access, debates continue about whether water agencies should promote gender equality through water projects because of a commitment to equity, or whether they should try to promote positive changes in gender relations to ensure that water projects provide more water. Increasingly the commitment of many agencies is to rights and social justice, and not just to expanding water availability (WaterAid website). However, this commitment exists within a framework of international water policies, which are contradictory (see Chapter 1). For instance, there is a commitment to increasing access for all as well as a commitment to the privatization of water as a way of increasing coverage in contexts of reduced government spending. Yet the private sector is not bound by any social commitments to the poorest nor to ensuring equity, and indeed is increasingly wary of undertaking such work: 'All dried up: As companies go cold over providing water to poor countries, the fear is that the public sector remains an inadequate solution' (J. Vidal, *Guardian*, 19 March 2003). While there is commitment to water for the very poorest, there is also a commitment to cost recovery, even though research into cost recovery, in health and education for example, has shown that it excludes the poorest from access (DFID 2004, WHO/UNICEF 2004a).

The inherent contradictions in policy followed the acceptance of water as both a social and an economic good in the Dublin principles (Sotanes and Gonzalez Villareal 1999): it is defined as a right (UN 2002) but it is also something to be paid for to ensure sustainability in contexts where poverty levels are high (World Water Forum 2003). The clear bias, as seen in the monitoring of the Millennium Development Goal (MDG) on water (WHO/UNICEF Joint Monitoring report 2004b), is towards increasing coverage and this is assessed very broadly. There are, as yet, no disaggregated statistics to show who is able to access and benefit from the increased water supplies across the various regions of many countries, nor is data collected below the household level.

The wider development context is currently one in which gender is not actively shaping most of the MDGs and their interpretation; gender work based on the Beijing Platform for Action (1995) remains largely isolated from these targets, around which so much development aid and political will is focused (Painter 2004). The earlier report on *Progress Towards Gender Equality in the Perspective of Beijing Plus 5* (OECD 1995) highlighted the limited progress on gender in bilateral agencies, something which continues as governments continue to debate gender equality strategies.[2] These debates about gender in development are mirrored in the NGO sector; gains made in the 1980s and 1990s seem to have been eroded globally in development forums and agencies. While the rhetoric usually remains intact, in practice gender is highly contested, often misunderstood, and a concept to which many are hostile; it is increasingly addressed in the wider context of 'vulnerability' or 'diversity', and sometimes ignored altogether.

It is within this fluid and evolving global policy context that NGOs try to address the technological requirements for improving supplies, where water sources are often variable and erratic and the geology is complex. They also try to engage local communities to manage and sustain these, to ensure that the water provided is appropriate, maintained and used in ways that benefit all. NGOs have paid far less attention, to date, to the organizational implications of taking gender seriously in water provision. This chapter takes one NGO as a case study to explore these issues in practice, to explore how international development agencies try to balance the institutional issues as well as the aims and objectives around technology and social, especially gender, relations. The specifics apply to this one agency, but the findings apply broadly to government and non-government agencies working on community-based water provision.[3] The NGO presented has gone further than many in taking a community development and gendered approach to infrastructure development and the delivery of water to poor people. This is an ongoing challenge requiring that the blueprints for service provision from the past, that so often excluded gender and poverty issues, are replaced with very new ways of working.

The chapter focuses largely on the work of the headquarters in the UK. The experiences on the ground, where these policies are implemented, are discussed in case studies in this book (see Chapters 11 and 12).

The Case of WaterAid

The Early Years

In the early 1980s, the water authorities in a number of countries[4] took up the UN challenge of providing domestic community water supplies in rural areas in developing countries. The UK water authorities responded by establishing a charitable trust in 1981 that became a company limited by guarantee in 1985; this has now become the largest UK NGO working on water supply, operating in fifteen countries in Asia and Africa. These countries have some of the highest levels of poverty in the world, with large numbers of people without good drinking water or basic sanitation services.

As an industry, the UK water sector evolved out of a historical experience that did not require attention to gender and poverty in order to ensure access to water and sanitation services by all. It was an industry staffed by engineers, as was WaterAid (WA), and dominated by a technical focus. These origins have played a part in shaping the culture of WA and the early policies it adopted. UK water authority staff formed the early steering committee and board of trustees, and it is difficult to assess how much consideration they gave to the possibility that a different approach might be needed to develop sustainable community water supply systems in poor countries.

The industry has always raised significant funds for WA, from water authority employees and water users, through local fund-raising events as well as through the water billing system. In 1983 they raised £300,000 rising to £1 million by 1987. The UK water industry was privatized in 1989 and the new private companies (PLCs) pledged to continue support. Leaders from the industry continue to be agency trustees and these PLCs undertake successful fund-raising for WA.[5]

Initially, the emphasis of WA was on designing sound technical, low-cost options to supply safe water for domestic use; it was not involved in irrigation or water for production, or the issues of long-term water creation and conservation.[6] The focus was on bringing existing water to poor communities for drinking and household use. The technical quality of the work was often high and the work was usually done in partnership with donors and local agencies, including local governments and NGOs. Engineer advisers, predominantly men from the UK, carried out in-country support; some were volunteers from the water authorities and others were employed staff. Gradually in-country offices were established and staff size grew, with the first regional programme manager being appointed in 1992.

The goal was to provide drinking water to as many people as possible, providing technically sound systems and good quality water supplies. They needed to be built quickly and coverage was tracked; success was measured by the number of water points installed. However, an external evaluation of one large WA project found that 'WA did not establish a sustainable community based operation and maintenance system, nor ways of assessing these' (Fawcett et al. 1995: 1). It lacked systems for measuring the sustainability of supplies or the impact on hygiene and sanitation, and this review found one third of supplies were not operational after a few years.

The Growing Internal Debate on Gender and Community Participation

By the early 1990s, the focus was shifting. The technical package was refined to include an increasing amount of training for community water committees who were to support installation efforts and on-going maintenance. Hygiene and sanitation education activities expanded and a policy was introduced in 1996 requiring all projects to integrate hygiene activities into their work. Discussions about gender and the issues of unequal social relations started in 1994 (later than in some other agencies), showing an increasing recognition of the need for flexibility in dealing with different social realities and complex gender relations in diverse contexts.

Several factors contributed to these changes, especially growing questions internally about the long-term impact of the new water systems, and whether water education programmes were actually leading to better hygiene and health. These concerns grew as more evaluations questioned the effectiveness and sustainability of the agency's work. External drivers included pressure from bilateral and multilateral donors to include gender in all development work, and the global commitment to bring water to all, including the huge numbers of poor urban and rural people in developing countries still without sustainable water and sanitation services.

These internal and external debates were informed by wider research from across the sector. Studies showed that a technical or 'hardware' approach to water and sanitation provision was inadequate for ensuring success (Black 1998). The breakdown rates in rural areas were high and the adoption of hygiene education messages low (Black 1998: 14). The challenges that had to be overcome to ensure maintenance and positive health outcomes were becoming better understood, as was the nature of the shrinking supply of potable water resources. The importance of using a gender and participation approach to ensure poor people had access to services was recognized (Black 1998: 44). Simultaneously donors continued to press for rapidly increased water coverage: 'The UN estimates that 2.7 billion people will face water scarcity by 2025. Some 40% of the world's population now live in countries with water shortages... Many say that there is no way that the UN target of halving the number of people without access to water can be achieved' (*Guardian*, 19 March 2003: 8).

The WA head office, influenced by and contributing to these debates, commissioned an increasing number of evaluations from the late 1990s, to ensure gender and poverty were addressed. Gender was made a 'big issue' in 1999, encouraging debate across the agency.

The evidence gained from country reviews and evaluations was mixed: while they often showed that the technical quality of the work was good – something of which WA was rightly proud and no mean achievement – the social impact and the distribution of expected benefits from the water supplies were not always positive. Some of the key findings were that:

- Men in communities made most of the decisions on project implementation and received the greatest opportunities for training and paid work.
- Women tended to do much of the unpaid work during construction as well as the cleaning and some maintenance of water points over the long term.
- The way the work was carried out tended to maintain existing unequal social relations – gender, caste, and class – within the communities, ensuring the domination and control of resources, including water, usually by men of wealth, high caste or class.
- Women's involvement in committees was there, but seldom meaningful as decisions were still predominately made by male members.

- The amount of time women spent collecting water was reduced, but this time did not necessarily enable women to improve their conditions or address their position in the community.

These findings highlighted the problems of meeting women's needs and led WA to ask countries to develop context-sensitive gender policies with their partners, although no guidelines were provided. Some consultants were asked to comment on gender issues in a few countries (see Chapter 6), though this was not common and follow-up money for new gender-based activities was often not available.

Another response to these findings was to commission a gender review in early 2000. WA wanted to develop a definition and understanding of gender that had meaning for the water sector, and to find appropriate tools for staff to use in implementing a gender approach. It was looking for clarity and answers to the difficult and thorny issues of how to work to address gender inequality while delivering improved services to people in real need.

This review did not lead to positive outcomes for WA's work on gender. While WA had expectations of learning about gender definitions and gender tools and frameworks that could be applied in the field, the consultants (also the authors of this paper) felt that addressing gender involved learning from experience and using that learning to inform changes to their work. They asked staff to analyse the specificities of gender relations in the different contexts where they worked, and to explore the impact of WA work in diverse communities, to identify successes and failures and how such work might be done in future. They did not think that gender approaches could simply be taught and grafted into WA's work. Staff found this approach frustrating. They asked instead that existing definitions, tools and packages be shared so they could select the most appropriate ones for working on gender in water projects; they wanted immediate solutions about how to work with gender issues.

They sought out other consultants (Development Planning Unit, London University) to provide them with these, but again this resulted in some tensions and frustrations. WA wanted established gender definitions and approaches that it could adapt and use for training alongside its existing packages for technical water provision, community development and hygiene work. However, gender practitioners, have learned the importance of developing an understanding of gender over time, based on experience, acknowledging its complexity and the importance of context, and that it has to be worked with carefully, over time, in different ways with different communities. More technical, 'short cut' approaches to gender mainstreaming often result in it being ignored or misunderstood.

The overall feedback to WA of the two gender reviews in UK in 2000/1 was:

In relation to programmes:

- The understanding and approaches, within WA, related to gender and social issues were wide.
- Thinking and practice varied considerably across partners, countries and offices on gender and social issues; specific gender objectives were not defined in the agency or in country strategies.
- A few countries were exceptional in doing work to understand and agree social objectives in relation to community-based water supply and sanitation work. They employed trained staff, partners and local people to work on gender within a poverty remit.
- Countries engaging in the social aspects of the work were setting differential tariff and subsidy rates to ensure that poor people could have access to water over long periods of time.
- Many partners and countries found the concept of gender vague and confusing. Their focus was primarily on delivery of high-quality water systems and some hygiene and sanitation education targeted to women. Their knowledge of how water and sanitation provision impacted on marginalized men and women was limited.
- There was growing interest among some partners and staff for WA to clarify its position on gender in relation to water and sanitation work. However, they wanted 'ready-made' principles to guide future work (Wallace 2000).

In relation to the organization itself:

- Women in the organization still found the organizational culture to be male and engineer orientated.
- Men dominated senior staff positions, including within partner agencies; women were mostly in middle-level posts.
- An agreed rationale for, or commitment to, a gender approach within the agency was not in place. It was unclear how far the organization at HQ level was engaging with gender work only because of its potential to increase funding.
- There was resistance towards Equal Opportunities at HQ level, though gender mainstreaming in the field programmes was more readily acceptable.
- 'Gender information that appears in evaluations does not automatically influence the design of follow-up activities... The agency focuses mainly on the impact of water problems and interventions on women ... but not necessarily on gender relations. Gender largely means women (and children)' (Levy 2001: 8).

- Some trustees, and senior staff, felt that gender would divert WA's focus from water. Engaging with the complexities of culture and social norms to address gender inequalities was felt by many to be beyond the remit of an agency, which aims to deliver good quality clean water to as many people as possible with limited resources (Levy 2001).

'Looking Back'

In addition to employing gender consultants, WA was keen to learn from its own experience and commissioned a twelve-month internal review in 1999 to explore widely the impact of its water and sanitation work. This was a participatory review, led by WEDC, a UK university water department. It involved staff and communities looking at the impact of the improved water supplies in Ethiopia, Ghana, India and Tanzania. It explored five hypotheses, one of which involved gender; this was that women and children would benefit more than men from better water provision.

Many of the findings were very valuable for WA, and highlighted that community management was essential for ensuring the positive impact and sustainability of water supplies, that communities needed considerable on-going support, and they were diverse and not homogeneous. Some other development agencies questioned aspects of the review at a feedback workshop on the preliminary findings. While they welcomed the retrospective nature of the study, and the focus on community perceptions, they questioned why WA had not built on existing extensive knowledge and experience of participatory gender and poverty work available in the UK NGO development sector (Welbourn 1995, Guijt 1996). WA had prioritized expertise from the water sector, above experience in participatory approaches to service delivery, which resulted in some complex social issues being raised that were hard for the WA team to interpret. Consequently, 'the study teams encountered difficulties in recording changes in gender roles following water and sanitation interventions' (WaterAid 2001: 22). They found that while women saved time on water collection their workloads had not reduced and 'in comparison, men have more time to relax than women' (WaterAid 2001: 23). While they found examples of women's improvements in income, health, and personal hygiene, these findings were not quantified. The review found 'it was difficult to validate ... [their hypothesis on women]... Across all communities, the women and children were typically groups who were most disadvantaged and hence they tended to benefit more from project interventions' (WaterAid 2001: 25). WEDC lacked expertise in this work, and many felt that WA's analysis of communities and gender could have been greatly extended if it had started from a more informed base.

The study was much clearer about what was happening with the water supplies, which were flowing well after several years and were a major achievement. This was because of the 'continued and on-going support to communities [which] facilitated sustained development ... and ... increased the impact of interventions' (ibid: 26). The few that had failed in Ethiopia had failed because of a lack of adequate community structures to maintain the systems.

WaterAid's Response to its Growing Understanding of Gender

While the evidence about the problems caused by a lack of gender analysis and strategies for tackling gender relations were clear from these different processes, the response was ambiguous. In the strategic plan written for 2000–5 the focus was on increasing the scope and coverage of WA; the work was to grow significantly and water supply targets dominated. Social issues were highlighted, but were to be addressed within that growth agenda. The mission statement said that WA was 'dedicated to the provision of safe water, sanitation and hygiene education to the worlds poorest people'. A commitment to meet the needs of women and children was stated, but while there was a statement to 'seek a fair gender balance in its decision-making processes' there was no explicit commitment about how to address gender issues in the work on the ground. There was little analysis of how fast it is possible to install water supplies within communities if people are to really participate and ensure issues of gender inequality are addressed. Targets were set for financial growth and greatly increased water provision for the largest number of people, but not for equitable access and benefits, or sustainability. While WA says that sustainability rests largely with the primary users, usually women and children (WaterAid 2004: 28), this was not clearly reflected in the strategy.

The new strategy, 2005–10, focuses on the needs of 'poor people', and 'the poor and marginalized'. Vulnerability and exclusion, caused by a wide range of factors – age, ethnicity, religion, social status, and HIV/AIDS – are highlighted. Gender is one factor of social discrimination but now has to be addressed alongside this range of issues. WA has adopted the current development approach of looking at diversity and the range of factors affecting access, defining gender as one of these. This is probably because, in practice, gender has become equated with women, and the complexity of gender theory which highlights the cross-cutting nature of gender with all other aspects of social difference has been overlooked (see Chapter 8).

Gender is specifically mentioned under monitoring and evaluation and programmes are asked to provide disaggregated data showing the impact of projects on different gender, age and social groups. Countries are also expected to show an understanding of how well services recognize and meet the different needs of men, women, girls and boys, although the priorities set for measuring results do not specifically include gender. The indicators focus on organizational learning, advocacy, relationships with major sector actors, effective monitoring and evaluation systems, evidence about how major water and sanitation sector issues affect the poor and marginalized (which could include gender issues), and increased coverage and financial growth.

The debates around gender and the social aspects of water provision that emerged during the work with different gender consultants continue in WA today. There is no agreement, in practice, about how far WA should get involved in complex and culturally sensitive community relations. While it is understood that communities are pivotal to the success and longevity of projects, it is less clear how much 'social engineering' within communities WA should engage in. Gender has never been a mandatory part of WA's work, in contrast to the clear policies and manuals on involving the community, working on health and hygiene, and technical aspects. There is no long-term staff post for gender work in London, and country offices make their own decisions about posts. A gender policy has yet to be written. Allocating money for developing the skills needed to deal with complex social issues in the context of technical and community programmes is not yet a priority. Training and space for staff to debate these difficult concepts and decide the way forward are limited, and there is currently resistance to the promotion of an equal opportunities approach. The understanding and application of gender concepts remain widely variable across the agency.

Senior management continues to disagree with the gender consultants' view that the aims of increased coverage and the fast expansion of water supplies are potentially in conflict with working with communities in ways which can uncover and start to address the social and gender inequalities that affect who does and does not benefit from new water supplies. Gender remains essentially an issue of choice. While staff are encouraged in different ways to take gender into account, the posts, policies, resources and guidance to support this work are limited.

Most reviews and evaluations continue to be undertaken at the end of projects, assessing short-term rather than long-term results, although encouragement and some guidance have been given to country offices about how to undertake long-term reviews. Evidence about how well water supply systems are maintained, whether they continue to operate at optimal supply levels, whether access and use are equitable, and how women do or do not benefit over the long term remains anecdotal and poorly recorded. Recent major reviews on management and learning undertaken from London did not include a gender perspective.

WaterAid, in line with many donors and other development agencies, has followed the UN declaration of 2002 defining access to water as a human right. It recognizes that 'water for all' is a very long-term goal, but it is a right, which requires a commitment to providing sufficient, affordable, and safe water to as many people as possible. The gender issues involved in ensuring women as well as men have this right in reality are not yet analysed on its website.

Middle Managers Keep Pushing the Gender Agenda Forward

The concern with gender was always identified with a small number of middle managers in WA, mostly women in HQ and the field. In 2003, a group of women in WA promoted the development of a gender resource pack, which was sent to all offices in 2004. This represents WA's first major attempt to record its experience, and that of others, in applying a gender approach to water and sanitation work. The resource pack builds on the gender work that different staff and consultants, mainly women, have undertaken in recent years. It provides specific examples of gender issues confronting WA and partners when working on the ground, and the implications of these for WA's work are discussed. Guidance on developing a gender monitoring and evaluation system for water and sanitation programmes is offered. The examples are all drawn from a few countries at this stage, but the intention is to promote contributions and learning from many more. This is starting to happen with interesting gender experience being offered from Ethiopia (on recruitment, research and policy dialogues for example) and West Africa, where gender guidelines and workshops are now underway in Ghana and Nigeria.

The pack provides WA with an important basis on which to promote discussions on gender at a very practical level. It grapples with the specificities of gender relations in different contexts and discusses real experiences. It is a good way to raise the profile and understanding of gender, based on a good internal analysis of WA's need for clear explanations and concrete examples to move forward. The pack does not – as yet – define a clear position on gender or draw together the other sources of learning about the social consequences of WA's work already available from past reviews.

Marrying Cost Recovery and Other Policies with Gender and Rights Approaches

WA supports a range of policies around water, in addition to those on access, rights and community involvement. It is especially known for its lobbying work on community involvement and participation in water projects. Some of its policies may not sit as easily alongside such commitments. For example, WA supports work with the private sector in water provision, as indeed do the Global Water Forum and all major donors. Some observers have deep concerns about this (Aegisson 2002), especially because this sector is not regulated in relation to social impact and equitable access, unlike government and NGO providers. Water companies are set up for profit maximization, not social justice.

WA is committed to the principle of cost recovery and works in ways that support the principle that recurrent costs are to be met by communities to ensure there is funding for long-term maintenance of supplies. Its practice – and that of its partners – varies widely, though the concept of cost recovery being key to long-term sustainability and community ownership is widely in evidence. WA has no explicit policy on subsidies or free access to the poorest; however, it has written papers exploring the issues around cost recovery and the role of the private sector, recognizing there are potential contradictions between these approaches and other values and aims it espouses (Gutierrez et al. 2003), and some WA partners do work on variable charges and water subsidies for the poorest. While WA challenges many of the big global players on issues of community involvement, local ownership and the need for long-term sustainability, it has not been so vocal in relation to gender.

There is little analysis in its documentation exploring what cost recovery means for women in different communities. Yet women often lack decision-making power over financial resources, and while they may be responsible for providing drinking water in a family, they often lack the leverage needed to get income to pay for it. A recent WA paper from Bangladesh noted that 'Women's ideas and opinions are very important re issues like the installation of water and sanitation project hardware components, but they are often ignored even in simple daily issues like the purchase of soap' (Ahmed 2002: 6). The extent to which charging for water discriminates against the poor, especially poor women, is explored in some of the case studies (Chapters 11 and 12).

The Country Programmes

A gender approach to the work is evident in some WA programmes – at the time of writing especially in Bangladesh, India, Nepal, and Tanzania – and steps to analyse gender within their work are underway within some offices and among some partners. Often these countries have particularly benefited from a champion or series of champions to drive the debate forward and to support gender analysis on the ground.

Work undertaken by WA country offices and their partners to develop a more gendered approach includes:

- Research done by staff to learn how the work done is affecting men, women, rich and poor.
- Case study reports by staff to document their experiences of working with men and women.

- Workshops to train staff and partners on gender, as well as to discuss findings from the research and case study write-ups.
- Discussion of gender at each monthly and other staff and partner meetings (keeping it on the agenda).
- Allocation of funds for gender workshops, training and pilot projects that take a more gendered approach to the work.
- Setting up of gender disaggregated reporting systems.
- Making explicit a commitment to gender in country strategies.
- Work to develop gender-sensitive monitoring and evaluation systems.
- Development and use of gender-sensitive recruitment processes.

In all these cases, a commitment by leaders (in WA or partner organizations) to develop a context specific gender approach is evident. Support to managers and field staff taking this difficult work forward is key to progress, especially in the absence of practical written guidance from the agency.

Countries working on gender recognize that their work largely follows a women in development rather than a gender and development approach. Discussion focuses on how women can participate, attend meetings and benefit from the organization's water provision projects. Most evaluations count the number of women participating at each stage of a project. These countries recognize the need to find an approach which will ensure the longevity of water supplies and sanitation services, and ensure accessibility for poor men and women equitably. In future they may analyse women's opportunities and constraints, how their roles are constructed and defined, and where their subordination is created and reproduced. This will help to identify where there is room for positive change in gender relations around water control, access and use.

Other countries say they want guidance on gender, a 'how do we do it guide. They don't know what the first step is so they don't take it.' 'Every country and partner ... (has) ... to develop their own methodologies. This was time consuming and challenging but made processes very area specific' (WaterAid 2004: 55). While Headquarters has asked for a country office to lead the debate and develop an agency policy on gender, no country has yet stepped forward. The provision of technically sound water supply systems remains the top priority of most in-country staff. Addressing the social complexities to ensure water systems are equitable and rights are realized in practice is a challenge that staff in most countries have not yet taken on.

Senior management teams remain dominated by men, although efforts are being made in several countries to improve the gender balance at senior management level and to encourage partners to do the same. All countries are asked to promote a gender balance in staffing, but this is challenging, because, for example, it is difficult in Bangladesh to recruit women at management level ... well qualified women are less in number in the market than men; due to the traditional reproductive and domestic responsibilities of women, some women can not cope with a professional job; still in Bangladesh women feel insecure to commute or travel alone; sometimes family members discourage women to work outside of the home. (Ahmed 2002: 8)

While some countries struggle with these shifts, the changes in the gender balance of staff are not yet monitored globally.

WaterAid's Gender Approach to Equitable Water Provision

The case study reveals an ambivalence about the role and significance of gender relations in the delivery of clean domestic water, even though the role of women has been increasingly recognized as important at every stage in the process from design through to water use and maintenance. The flurry of interest in gender at the end of the 1990s was not followed through so while WA still encourages staff to think about and address gender issues, doing so is largely optional and up to the individual. When it comes to the allocation of resources, other priorities prevail. There is overt commitment to water as a right, implying equitable access and the need to overcome barriers for the vulnerable, but how and in what ways gender affects these aspirations remain undefined, both at policy and implementation levels.

Conclusions on the Factors Shaping WaterAid's Response to Gender

Factors Against Taking Gender as a Core Principle

The UK Water Industry   Links to the UK water industry remain strong through funding and governance, with over 60 per cent of trustees connected closely to the UK water authorities. Most of the privatized water PLCs are members of the organization as are the major trade unions and research centres associated with the water industry. The issues around how poor people access water are not very familiar to this constituency, which is committed to meeting targets, and ensuring the high demands of technical provision and maintenance of water.

There even appears to be a concern among some trustees that engaging too deeply with complex social realities might detract WA from its core business. In addition, these issues are of limited interest to many core WA supporters in the UK. WA sees much of its success as meeting its targets, delivering on time, and communicating clearly with its supporters.[7] Increasingly it is the case that UK NGOs find it hard to engage with the complexity of social structures and inequalities in their public communications, even though these do affect the success of development work long-term. The formula for communicating with the industry and the wider public for raising funds, which is proven and effective, tends to rely on clear messages about bringing water to people in need.

Senior Management Commitment Middle managers are predominantly women in WA, and appear to be more committed to working on gender than senior management. The leadership response has been ad hoc and no serious resources have been invested in this area to provide support. While no one blocks staff or country programmes from gender work, the agency has not invested sufficiently to ensure effective action. Senior management remains predominantly male, although recently more women have been appointed to director posts in the field (five at the time of writing). Only four women sit on a board of sixteen trustees and while being female does not equate with being committed to gender equity, in most agencies – including WA – women are the key drivers of change around unequal gender relations, within organizations and programmes.

The significance of commitment from senior management is well illustrated by reference to two issues: financial growth and recruiting staff from southern countries. The 2000 strategy committed WA to greatly increasing income; that target was met in the next four years. WA more than doubled its income in a few years, from an average of £9 million at the end of the 1990s to around £20 million by 2003, at a time when many small and medium UK NGOs were struggling to survive and few grew significantly.

Similarly, the new director had a passionate commitment to changing the profile of the organization to an international one. Over a short period he has shifted the management from predominantly white males to a situation where eight out of fifteen directors abroad are from the South, and Southern staff now largely fill the regional manager posts in the UK. Two Southerners have also recently joined the board of trustees. This change in the number of non-UK staff in leadership posts in a short time demonstrates clearly that when there is strong, consistent commitment from senior leadership within an organization change does take place (Goetz 1998).

The Desire for Greatly Increased Water Coverage By 1993 the agency reported that it had provided safe water to 1.8 million people. By 1997, the figure had reached 3.5 million, and by 2001 6.5 million. Annual coverage estimates are 500,000 people. The agency has set an aspirational goal in its new strategy to reach a total of 10 million people by 2010. It is expected that costs per beneficiary will be tracked and such costs will be kept low.

This commitment may clash with the commitment to a rights-based approach, where women as well as men have rights to water. In order to meet these coverage targets, programme work in most communities will still have to be done within a one-year time frame, though WA country programmes which have undertaken work on gender and poverty issues talk of doubling this time frame. A longer time frame is undoubtedly needed to ensure participation by marginalized groups, especially women, takes place, and to enable local people to address the issues of inequality as they arise (Porter et al. 1999). This will conflict with the speed of progress implied by the new targets for WA.

The Need for Clear Demonstrable Results   Official funding in particular brings with it pressures to demonstrate clear results within specified time frames, which are usually quite short. For instance, under its grant agreement with DFID, the agency is expected to share the successes of WaterAid's community-focused approach. DFID wants to learn how the approach can be scaled up and improved for use at national and regional levels. Building in a gender-sensitive approach to water and sanitation work is not easy; besides time, it needs resources, commitment and effort by everyone across an agency. These requirements and WA's ambivalence towards gender could make it difficult for WA to develop a gender and poverty approach to water and sanitation provision and convince major donors that this could be effective in terms of improving equity and the long-term viability of infrastructure on the ground.

Most agencies in the UK feel beholden to their donors and find changing the parameters of grants and agreements difficult. Unless there is a real commitment to ensuring gender is central to those agreements it often drops off the agenda. This has been the case recently where concerns with coverage and cost targets, and efficiency and effectiveness seem to have eclipsed earlier donor commitments to issues such as gender and the environment.

Staff Resistance   Most development agencies experience staff resistance to working with gender issues (Porter et al. 1999). Gender issues affect staff personally as well as the organization and its programmes. These issues are complex, demanding, and time consuming. Many staff feel that taking them on goes beyond the remit of the agency and interferes in local cultures. Many staff in WA prefer to work within the approaches WA has developed for building water supply systems, using appropriate technology and involving the community in ways that ensure that the water points are built and fees are collected to ensure long-term maintenance. Experience from other organizations shows that only a strong organizational position on difficult social issues, grounded in experience, can promote change and a willingness to engage in challenging areas of work.

A Contradictory Policy Environment   Development agencies involved in water provision are working with a range of concepts, approaches, policies and targets that are often assumed to be compatible, but are in fact pulling in different directions. WA is aware of the tensions and has explored them in different research documents. Nevertheless, it continues to grapple with meeting cost recovery demands while trying to reach the most vulnerable people; with increasing coverage while trying to ensure community involvement; and with working with water as a public and social good, as well as a private resource and economic good. The varying policies and priorities set by donors, the water sector and WA itself continue to drown out the calls of some for a commitment to gender as an essential part of ensuring the provision of sustainable water for all.

Factors that Support Work on Gender within WA

The countervailing pressures undoubtedly come from individual committed staff – largely women – based in the UK and field offices. Questions asked by country staff, partners and some donors, and the findings cited in evaluations have all helped to move the discussion on. Gender is included in some work plans and on regional workshop agendas, and individual staff have worked hard to promote the issue on the ground in some countries.

There is evidence of learning from reviews, though at the moment this learning remains scattered. Some documentation is powerful and clearly describes the effects the work on water projects has on women's workloads and the barriers that women face in implementing the health practices promoted by hygiene education programmes in specific cultural contexts. The difficulties poor people face in participating in decision-making processes around water and sanitation systems, paying for services and attending meetings are also highlighted. Much more could be achieved if all the learning was collected together and used to inform future planning.

The commitment to water as a right, with the equity issues implied in a human rights approach to development, may also become a force for more widespread work on gender in future. However, strong external drivers pushing for a gendered approach to development are fading. This will make it harder for those interested within the organization, as within other NGOs, to implement an approach to community water provision that ensures equitable access for all. Their success though is important as WaterAid is a widely respected NGO in government, donor and NGO circles so the work it does can be influential. It has internally debated a more gendered approach to community water provision and the group of gender champions has grown in headquarters and across a number of the country programmes. It is to be hoped that WaterAid will enable these champions to build on their knowledge, understanding and experience to develop an explicit gender approach, which can become embedded in future work.

Notes

1. This chapter is the responsibility of the authors alone. It was written following several consultations with WA staff, who have also commented on all the drafts. WA does not agree with every argument in this chapter, though staff who have commented agree that gender work has not been prioritized yet, in practice, within WA.
2. DFID currently does not have a gender equality policy; this is the subject of internal debate.
3. In the last ten years, many UK-based international NGOs (INGOs) have supported water projects in developing countries and have faced the same tensions and pressures explored in this case study. This includes most of the larger UK INGOs such as ActionAid, OXFAM and Save the Children.
4. Water industry created NGOs are Water for People in the USA, Water for Survival in New Zealand, Watercan in Canada and WA in the UK
5. Thames Water is one of the largest water companies in the world. Since its privatization in 1989 it has raised over £14 million for WA via its staff, customers, community groups and suppliers, who are all encouraged to support the agency in a number of ways.
6. WA does not yet work on issues around the creation, preservation or conservation of water; work on these issues is increasingly under threat because governments lack the necessary resources.
7. WA has a 'winning formula' for fundraising as one consultant put it. It has simple messages about providing water pipes and supplies to poor people, and its supporters are largely uninterested in the complexity of social inequalities or the challenges of reaching the poor in practice.

References

Aegisson, G. (2002), *The Great Water Robbery*, London: Standpoint, One World Action.

Ahmed, R. (2002), 'Integration of Gender Perspectives in WaterAid's Bangladesh Programme: For Improved Safe Water Supply, Environmental Sanitation and Hygiene Promotion', Dhaka: WaterAid. Unpublished discussion paper.

Black, M. (1998), *Learning What Works: A 20-Year Retrospective View on International Water and Sanitation Corporation 1978 to 1998*, Washington: UNDP, World Bank Water and Sanitation Programme.

Davies, J and G. Garvey (1993), *Developing and Managing Community Water Supplies*, Oxford: Oxfam.

DFID (2004), 'Accelerating Action on Girls' Education: Rights and Responsibilities', London: DFID. Unpublished.

Fawcett, B. L. Juppenlatz and J. White (1995), *WaterAid in Uganda: Busoga Region Borehole Rehabilitation Project: Uganda Evaluation*, London: ODA Evaluation Department.

Goetz, A.M. (1998), 'Getting Institutions Right for Women', *IDS bulletin*, 6(3).

Guijt, I. (1996), *Questions of Difference: PRA Gender and the Environment*, London: Institute of International Environment and Development.

Guijt, I. and M.K. Shah (1998), 'Waking up to Power, Conflict and Process,' in I. Guijt and M.K. Shah, eds, *The Myth of Community: Gender Issues in Participatory Development*, London: Intermediate Technology Publications.

Gutierrez, E., B. Calaguas, J. Green, and V. Roaf, (2003), *New Rules, New Roles: Does PSP Benefit the Poor?* London: WaterAid and Tearfund on WaterAid website.

Levy, C. (2001), Report on the gender workshops in WaterAid, London: WaterAid. Unpublished.

Mathew, B. (1991),' The Planner Manager's Guide to Third World Water Projects', in T. Wallace and C. March, eds, *Changing Perceptions; Writing on Gender and Development*, Oxford: Oxfam.

OECD (1995), *Progress Towards Gender Equality in the Perspective of Beijing +5: Beijing and the DAC Statement on Gender Equality*, Paris: OECD/DAC.

Painter, G. (2004), 'Gender, the MDGs and Women's Human Rights in the Context of the 2005 Review Process' Paper presented to the Gender and Development Network, London: Womankind. Unpublished.

Porter, F. I. Smyth and C. Sweetman (1999), *Gender Works: Oxfam's Experience in Policy and Practice*, Oxford: Oxfam.

Solanes, M. and F. Gonzalez-Villareal (1999), 'The Dublin Principles for Water as Reflected in a Comparative Assessment of Institutional and Legal Arrangements for Integrated Water Resources Management', Global Water Partnership, http://www.thewaterpage. com/SolanesDublin.html.

UN (2002), 'A Right to Water', http://www.wateraid.org.uk/in_depth/policy_ and_research/ right_to_water/default.asp.

Wallace, T. (June 2000), WaterAid Final Report on the Gender Review from December 1999 to May 2000, London: WaterAid. Unpublished.

WaterAid (2001), *Looking Back: The Long-Term Impact of Water and Sanitation Projects*, A condensed version of the WaterAid research report *Looking Back: Participatory Impact Assessment of Older Projects*, London: WaterAid.

—— (2004), Gender Resource Pack, London: WaterAid. Unpublished.

WaterAid website: www.wateraid.org.uk. Water Rights website: www.righttowater.org. uk.

Welbourn, A. (1995), *Stepping Stones: A Training Package*, London: strategies for Hope and ActionAid.

WHO/UNICEF (2004a), Joint Monitoring Report on Water MDG, Geneva: WHO.

—— (2004b), Joint Monitoring Report on Maternal Mortality, Geneva: WHO.

World Water Forum (2003), The 3rd World Water Forum, Kyoto, 16–23 March 2003, http://world.water-forum3.com.

8

Misunderstanding Gender in Water: Addressing or Reproducing Exclusion

Deepa Joshi

Introduction

This chapter argues that the current approach of the domestic water supply sector to gender is based on a misunderstanding of the concept of gender, which results, at best, in a failure to address exclusion and, at worst, in a reproduction of existing practices of exclusion. This has negative effects on the very people current water policies aim to assist – including the poorest women.

Gender theory stresses that gender is related to, but distinctly different from sex (Oakley 1972). Gender identities vary across cultures, 'every society uses biological sex as a criterion for the ascription of gender, but beyond this simple starting point, no two cultures would agree completely on what distinguishes one gender from the other' (Oakley 1972: 158), although universally patriarchy ensures structural and symbolic inequality between women and men (Molyneux 2001). However, race, caste, colour, religion and class deny a universality of gender inequality and women and men are not homogeneous categories (Mohanty 1991). Just as inequality by gender is never pure, it is never absent; rather it is reinforced across all levels of social organization (Whitehead 1979). In addressing gender inequality it is essential to pay attention to all these perspectives.

In translating gender theory into development policy and planning, there has been a reversal of thinking, from gender back to sex, 'separating and isolating women as the central problem from the context of social relations' and assuming gender inequalities exist only at the level of the household (Baden and Goetz 1998: 22). The reasons for this are political and technical; addressing inequality in gender relations involves understanding and challenging local customs and traditions and inequitable global policies and practices. It also requires time, and within the dominant development paradigm designed to achieve economic efficiency, it is possible only to pay lip service to these concerns because project targets have to be met. Involving women in projects, rather than challenging the social construction of gender inequality, is easier to achieve and is also quantifiable.

135

Field research into a popular water project, widely acclaimed for enabling communities to manage water delivery and address gender in rural villages in North India, revealed that involving users, especially women, to implement projects did contribute to project efficiency. This link has been reported elsewhere (Cleaver 1997). However, women benefited differentially and the most vulnerable gained little by way of access to water and/or furthering of equality in social relations, because the interpretation of gender in the project did not reflect the complexity of inequality within the communities. A singular, composite image of the generically 'poor third world women', much argued against in gender theory, was the focus and issues of caste and class were only loosely addressed. There was also little exploration of gender inequality in institutions, beyond the household and community.

Women's Movements in Evolving Development Policy

Oakley's contributions to the 1970s feminist movement in the North coincided with efforts to recognize women's multiple roles in development policy and planning. Hitherto women's contributions to production had been ignored; they had been seen largely as passive beneficiaries of trickle-down welfare development. This approach broadly involved organizing underdeveloped, backward nations in the South to adopt systems of knowledge and technology from the North in order to modernize; the model of the nuclear family with a male breadwinner, dependent female housewife and children was the basis of development planning (Kabeer 1994). It was assumed that by increasing opportunities for men, development benefits would trickle down to generically poor and needy households and to poor women as needy housewives.

Boserup (1970), citing the relative independence of women in the female-farming regions of sub-Saharan Africa and South-East Asia, challenged these assumptions of female domesticity and argued that women were crucial to production and so to development. This perspective became accepted and the 'Women in Development' (WID) approach was adopted in the 1970s. This shift coincided with the mounting debt crises in Africa especially, which led to structural adjustment reforms; these imposed economic liberalization on poor countries through the classic remedies of public sector cuts, and fiscal and export-oriented reforms (Elson 1991). Boserup's work on women's potential role in production found wide acceptance at a time when structural adjustments were biting and major Northern development agencies were quick to accept and integrate the Women in Development ideology in development aid policy (Kabeer 1994). Women presented an attractive and hitherto unused source of available and willing labour for efficient development (Davidson 1990).

Over time, women became identified as 'the fixers' of environmental problems in development projects (Jackson 1993). Prominent eco-feminists stressed women's special relationship to the environment (Mies and Shiva 1993), and described pre-colonial societies as harmonious and egalitarian, where women's livelihoods were supportive to nature renewing itself (Shiva, 1989). In their sustenance roles, especially those of fetching water, fodder and wood, women were identified as inherently respectful of nature and understanding of their environment. Eco-feminists contested the dominant model of development based on economic growth, with its emphasis on the role of men as providers. They argued that the environmental degradation brought about by the exploitation of natural resources for economic growth systematically depleted women's relationship with nature and therefore the resources for 'staying alive' (Shiva 1989). The fact that the roles of, for instance, fetching water and gathering fuel were not natural to women but were socially allocated was not analysed (Agarwal 1992, Leach and Green 1995) and 'environment' was added to women's already long list of caring roles. There was little consideration of the high opportunity costs of legitimizing the practice of assigning roles by gender in the context of depleting natural resources. The Women, Environment and Development (WED) approach tied women inextricably to the environment and this was picked up by the water supply sector – domestic water projects became women's projects (INSTRAW 1989). The eco-feminist questioning of the efficiency approach with its emphasis on economic growth was not taken further, and the development outcome was essentially a shift for women from welfare to an efficiency approach; equality was sidelined (Kabeer 1994).

Within the efficiency approach women's reproductive roles were not given the same economic value as their productive labour, and the focus on women's productive tasks did not require a transformation in masculine behaviour. Women, by taking responsibility for the task of reproduction, had liberated 'men to function as rational producers'; however, there was to be no such complementary support for women. The approach ignored the social roles of reproduction and the inequality of gender rights, and there was little clarity on how women on their own would become rational producers. Critics noted that the emphasis on women's productivity, in isolation, could backfire. There could be a backlash if it were proved that men had higher economic productivity than women, something that could be expected because of women's multiple roles and responsibilities (Jacquette 1982).

The approach also assumed that women's integration into economic production would provide women with an equal position to men; unequal power relations were not questioned. The issue of control was overlooked. Women were expected to join the labour force, but men continued to hold the rights to resources such as land and money (Rogers 1980). 'One cannot help concluding that the real issue is who controls the resources, not who does the work. Land, labour, livestock, capital, technology, information – are all valued goods that imbue those who own or control them with power and prestige' (Dixon 1985: 32).

Which Women?

The 1970s feminist protest in the North was critiqued by some for its 'whiteness' in the analysis of gender inequality. A new discussion explored the diverse range of women's struggles and political perspectives in the histories of different cultures and societies (Molyneux 2001). Questions were raised about the dominant image of 'the third world woman': 'the historical heterogeneities of the lives of women in the Third World ... [are] produced/represented as a composite, singular image of *third world woman* – an image, which appears arbitrarily constructed, but nevertheless carrying with it the authorizing signature of Western humanist discourse' (Mohanty 1991: 53). A network of African women (AAWORD) asked, 'What is this nature of development, from which we alone have been excluded and now need to be integrated? Have colonialism in the past and the asymmetrical world economy and political reality been so generous as to put all males in structurally dominant and skilled positions?' (Kabeer 1994: 33) The identification of class, race, wealth and religion and other social factors as contributing to social and economic inequalities significantly broadened the theoretical understanding of gender.

Towards a Complete Gender Equity

Early in the gender equity debate, Whitehead (1979) pointed out that organizations and their frameworks of rules and norms, commonly assumed to be gender neutral, in fact mirror society and the social order. They thus perpetuate exclusion by gender and reinforce differential opportunities of and access to education, skills, occupations and positions. Others realized that gender relations are subject to change as society changes; they cannot be universally assumed (Leach 1991). However, it was agreed that the relative powerlessness of women in most societies would not enable them to change their position on their own. Men and women would need to work together to address and counter gender imbalances, especially within the institutions that had historically upheld the subordination of women.

These insights led to the establishment of Gender and Development theory, according to which addressing gender would require understanding gender relations and how inequalities are reproduced in different contexts. Finding ways to enable dispossessed groups to change their position in society and set their own agenda of development became critical (Moser 1993). Taking the water sector as an example, the section below shows how this analysis and understanding of gender failed, for various reasons, to get translated into practice.

Whither Gender in Water Supplies Programming?

In the 1950s and 1960s the development of water supplies focused on expanding coverage by establishing centralised water supply institutions, installing infrastructure and promoting improved health. Women's water roles were not considered at all until the 1970s and 1980s. Subsequently, the focus was on:

- Increasing coverage
- Identifying water as an economic good, a finite environmental resource
- Increasing project efficiency through women's involvement
- Involving women in water management – based on assumptions about women's links to water and the potential rewards for them: improved access to drinking water, better family health and time to pursue economic goals
- Improving the position of women: structural improvement of their position through water management roles was anticipated.

What Benefits were Real for Women?

Women's involvement in water projects in the 1970s and 1980s secured project efficiency/success in some contexts, but often at the cost of women's voluntary involvement. However, improved water delivery systems were often not sustained: long queues for hand-pumps and the failure of supply to taps and tanks were common. Further, water delivery systems often focused on safe drinking water, and overlooked women's multiple household water needs, pushing them to continue to use traditional sources for washing. Kamminga (1991) reported that, even where there were realistic improvements in supply, time saved collecting water may have been reallocated to attending meetings, collecting user fees, cleaning water points and operating and maintaining hand-pumps and wells.

Women's involvement was often restricted to women becoming hand pump care-takers, health educators and users of waste water for growing kitchen gardens. There was silence about challenging the hegemony of gender roles to ensure that men shared water burdens equally (Chibuye 1996). Scant attention was paid to whether different women had the resources or the institutional support for effective social involvement in previously male-dominated domains, let alone the many requirements needed for gainful production or improving social status (Leach 1991). Projects especially marginalized the concerns of poor women by overlooking the differences between women. Attempts to address the many barriers facing women were largely missing, or were dismally shallow and apolitical. The overriding focus on involving women at community level masked the power differentials and political interests within water institutions, which shape water management and use (Mehta 2000). There was little information on how water organizations operated, or how they organized women's involvement within them (Rodda 1994).

The 1990s – Rhetoric of Gender in Water

New approaches, decentralization and community participation, and demand-management, were introduced in the 1990s and define current domestic water policies. However, the framework for water management remains over-archingly focused on achieving efficiency, though positive links assumed between equity and efficiency are yet to be proved (Mehta 2000).

Decentralization and community management were promoted in part because of World Bank research, which showed user willingness to pay for water, and the strong links between this cost-recovery and a sense of increased ownership. Ownership is further linked to increased participation, which in turn is assumed to promote scheme sustainability: 'participation can lead to greater self-reliance of local communities; less capture of the good by elites and equitable access to improved water supplies' (Narayan in Prokopy 2004: 3).

The Demand Responsive Approach (DRA) hinges on both community participation and decentralization, recognizing that rich men and women and poor men and women may want different kinds of service. From being historically supply led, the water sector is encouraged to be responsive to users' demands/ needs and allow user choices to guide key investment decisions, ensuring that services provide what people want and are willing to pay for. In exchange for contributions in cash or kind, stakeholders are ensured a voice and choice in technology type, service level, service provider and management/financing arrangements (Dayal et al. 2000). Government roles change from that of provider to facilitator of community and/or private sector involvement.

Prokopy (2004) identifies research that challenges the assumed links between participation and project success in the water sector. There is little empirical evidence to suggest that cost-recovery increases ownership or participation (Schouten and Moriarty 2003) or that participation necessarily has positive project outcomes (Cleaver 2001). This is especially the case where the conceptualisation of 'participation' is very broad. Voice and choice in most demand-led water projects are expressed by a representative committee and/or local leader/s, yet Sara and Katz (1998) found that demand-responsiveness is most effective when the demand is from user households; the level of effectiveness drops significantly when expressed through committees and local leaders. Analysing two popular water projects acclaimed for their success in user participation in India, Prokopy (2004) found only low to middle-level participation, that is, contribution of money, labour or materials in a pre-designed project, and/or participant consultation about largely predetermined issues.

Implications for Gender and Equity

Empowerment and equity are terms used liberally in current water policies, yet they have a very different meaning from the same terms used in gender theory. The following issues highlight this divergence:

- Community participation commonly ignores social divisions within communities and often results in more work for those with least power. Often, the poorest women of low social status are mobilized to undertake the voluntary, labour-intensive and socially unappealing tasks (Slayter in Jackson 1993). Participation may even be forced upon a community, taking away from residents valuable work and leisure time (Cooke and Kothari in Prokopy 2004).
- Concern for equity is contradicted by the DRA commitment to providing more or less water on the basis of ability to pay. Adequate evidence exists that for the poorest, free water is still the best water (Mehta 2000). In addition, Harrison (1993) reports that payment for water by the poor does not ensure access to safe, reliable and adequate water because multiple social variables determine access to water. The 'willingness' to pay argument may fail to address gender, caste and class inequalities, which all determine access to and control of water.
- The issue of whether women's willingness to pay translates realistically into ability to pay is not explored; neither is the issue of who makes decisions on behalf of households. The language of the DRA is gender neutral and ignores inequalities in resource allocation within and beyond households (Cleaver and Elson 1995); yet household systems often leave women with no control over financial resources.
- References to gender seem largely random; the term is now so loosely used that it can erroneously mean both 'women' and 'women and men'. Often, while women are recognized as the main providers and users of water (Guijt 1993), this also causes resentment, leading one co-ordinator of the gender session at the Hague 2nd World Water Forum to say, 'men will no longer have to grow long hair to resemble women in order to derive the benefits of water projects. Water projects would now meet the needs of men as well' (Visscher 2000).

This incoherence in policy illustrates the varied interpretations of gender currently in use in the water sector. It is against this backdrop that an analysis is made of one field reality.

The Village Setting: Heterogeneity and Disparity

Chapter 3 illustrated the links between gender, caste and water, which determine the water rights for women and men in small village communities in Uttaranchal

State in Northern India. Chuni village is similar socially and geographically to the neighbouring villages where the World Bank-funded water project, SWAJAL, was ongoing at the time of this research. It is therefore possible to use the findings from Chuni as a baseline for informing an understanding of the gender impacts of the SWAJAL project.

Location, Ownership and Access to Water Sources in Chuni Village

The key learning from Chuni was that:

- Water development and management systems are still largely traditional. The preferred source for drinking water is the traditional *naula* (natural spring), which is clear, clean and almost perennially available. In the absence of a *naula*, water is obtained from seasonal waterfalls (*gads*) during the monsoon, which often provide muddy water. When both are available, *gad* water is used for bathing, washing and feeding animals.
- Chuni is known as the village favoured by the water goddess, a water abundant village. However, the evidence showed that the stringent relations of hierarchy and power determine access to adequate water and those disadvantaged by caste, gender or economic class are deprived of water and excluded from decisions about water management.
- All the households who are members of the dominant caste group, by descent or marriage, benefit from adequate, appropriate water. Powerful individuals in the village, privileged by caste, class and gender, have historically held positions of decision-making in relation to water distribution and use; this practice continues. In contrast, the few water sources that were available to socially vulnerable groups and households in the past have been gradually taken away. Socially disadvantaged families are often forced to exchange safe, reliable and appropriate water for an unsafe, unreliable and inappropriate system and are powerless to challenge the illegality of the process.
- The men of the dominant caste and socially powerful families have historically violated official legislation around water and used their authority to secure official permission to divert water. They knew that the sanctioning authorities had little interest in coming to assess the situation and that most disadvantaged families would not dare protest.
- Changes in legislation providing official ownership of land for Dalits did not translate into corresponding rights to water for them.
- Women are more disadvantaged than men in their ability to influence water management decisions, due to notions of appropriate gendered spaces as well as the unequal allocation of reproductive work. The burden of water-related work, socially assigned to young women, is cemented on marriage.

• It is strikingly clear that in this village setting, some women are affected more severely than others by the lack of adequate and reliable water. Some women, by virtue of being the wives and daughters-in-law of powerful males, no longer have to carry water; the technology of 'cement tanks and pipes' lightened their water burdens. Yet the improved water supplies carried high negative costs to others and were most agonizing for the socially disadvantaged Dalit women.

The SWAJAL Project

The SWAJAL ('pure/own water') project delivers domestic water and sanitation schemes in rural areas in two states, Uttar Pradesh and Uttaranchal, in India. The project was planned and supported by the World Bank and managed by the Department of Rural Development, government of Uttar Pradesh (Uttaranchal was a later division of the state of Uttar Pradesh). Phase 1 spanned a period of six years starting in 1996, Phase 2 continues.

The project took a demand-responsive approach, based on underlying assumptions about community homogeneity and shared interests. Community participation was promoted by NGOs assumed to be receptive to the problems of the community and women in particular. They set up Village Water and Sanitation Committees (VWSCs), who select scheme type, collect user fees, purchase materials and arrange labour for construction. Women's participation was encouraged: the project aimed to enable women to take the lead in decision-making and share in collective benefits (World Bank 1996).

The goal of 'women's empowerment' translated into:

• 30 per cent participation of women in the VWSCs
• Women to become VWSC treasurers where possible
• Health and sanitation awareness training for women
• Non-formal education for women
• Income generation training for women; time saved from improved water supplies was to be translated to improved productivity.

Project guidelines did not recognize heterogeneity amongst women. It was assumed that 'women' would be involved in making decisions and gain equitably from other project benefits. The project specified 20 per cent representation of Dalits in the VWSCs, recognizing caste inequality in India; however, there was no specification about whether this meant Dalit men or women.

The success of the project is widely acclaimed and was instrumental in shifting India's rural domestic water policy from a supply-led to a demand-driven approach. This shift was also encouraged by reviews in the late 1990s which highlighted the poor financial and operational management of bureaucratic supply-driven approaches: 'the under-performance of the rural water supply sector is likely to continue unless there is a fundamental reform of the service arrangements' (World Bank 1998). World Bank-led research identified that it was becoming increasingly evident that the government alone would not be able to provide the necessary expansion of services to a growing population. These analyses were well accepted, and while the government of India accepts water as a social right, it also accepts the DRA, including cost recovery, based on a community development approach (Government of India 2003). Equity and gender are barely included in current government policy guidelines; there is no mention of 'poverty' and any references to gender are limited to 'involving women' in predetermined project activities (Joshi 2004).

The importance attached to addressing gender is more evident in World Bank publications, including those of the SWAJAL project. However, gender is equated to women and homogeneity amongst women is assumed. All rural women are defined as poor, and women's explicit links to water are assumed in project design and monitoring indicators. The project objectives define, 'a safe and reliable source of domestic water supply, a necessary precondition for improving health standards and economic productivity of (poor) women and men in Indian villages' (World Bank 1996: 1). The focus is on the successful completion of the project and the positive role of women in it, as seen in the brochures showing proud VWSC women members smiling; however, which women they represent from the heterogeneous communities, is not explained.

Project Outcomes on Social Relations of Inequality
The first village to implement the scheme was Mala in Uttaranchal. As identified locally (see below), Mala was not a water scarce village prior to the project. The communal stand posts built by the project certainly did increase convenience for some, especially those who lived close by, who were incidentally mostly members of the VWSC. These households in most cases had managed to divert water directly from the communal stand posts through detachable plastic pipes to their homes. In contrast, of the eight poorest households in Mala only three had a stand post close to their homes.

The open design of the tap stands meant that women still used the original (sheltered) sites for bathing and washing. Ironically, people said they preferred the spring water for drinking water, because of the smell and the visible white froth of chlorine (used to disinfect water in the storage tank) as well as the locally held beliefs about the 'rejuvenating' qualities of the spring water.

This case study focuses in particular on the impact of the project approach on gender and social equity, and the findings are supported by knowledge from Chuni. Analysing the project in Mala and several other local villages a year after completion revealed that despite its clear aims, the project had reinforced rather than addressed some of the gender disparities. This could have been expected, given the existing social structures in the area and the way gender was conceptualized in the SWAJAL project goals. A more recent study of twenty participating villages in the SWAJAL project (Prokopy 2004) reported somewhat different findings, including high gains in terms of equal and appropriate access to water. However, the data were not disaggregated by either gender or caste, and the research did not explore pre- and post-project conditions of water scarcity and access, issues that were identified as significant by local residents. A positive finding was that there is now good local decision-making around water, by user communities.

Participation and Decision-Making

The technical guidelines specified that 'eligible villages (should) have an un-disputed perennial source(s) in the village with water flow not less than 7 litres per minute as well as demonstrate community willingness and ability to cost-sharing' (World Bank 1996: 1). Village selection was left to NGO discretion and there was no consultation with local water agencies on coverage and access statistics. Mala was the stronghold of the local NGO and, in line with common practice, was a recipient of all projects allotted to the NGO. The gains were to be mutual: given the NGO's long history of working there, the chances of achieving project demands were reasonably high. In the view of one local field (NGO) officer, 'the project is not about providing water to needy villages and the rural poor. It is about proving right the demand-responsive approach.'

The distribution of services within the village remained inequitable, and those usually excluded continued to be so.

> In Hilay village, Hari Ram, a Dalit man nominated to the Committee, said, 'I was selected because Dalit representation was demanded. But my selection does not mean that I can challenge or influence anything done by the larger community. I only add Dalit colour to the Committee. Today, as you see, was a VWSC meeting. It was planned in conjunction with a ceremony in the house of the village teacher, who is "upper caste". I cannot attend the meeting in his house, as this is not socially acceptable. When we Dalits cannot sit and smoke the hookah together with the higher caste men, how can we plan together?'

There were no project indicators to measure the effectiveness of Dalit representa-tion in the VWSCs, or to assess gender in Dalit representation.

Women's involvement in the project was not always effective. Agarwal (1997) identified that, membership apart, to participate in decision-making women need to attend and be heard in public meetings. Yet eight months after her nomination as a member, Bheema Devi of Nachuni village was clueless about her committee roles and responsibilities. As a pregnant young daughter-in-law, she was unable to attend meetings let alone discuss decisions. Sheela Devi, being literate, was an appropriate choice as the Nanchuni VWSC treasurer. However, with her husband away in the city, she struggled to manage her fields, animals and five children. She was too busy to collect community contributions, deposit the money in the nearby town bank, keep records, make purchases for construction work or pay the labourers – all tasks demanded voluntarily of a VWSC treasurer, a role especially recommended for literate women.

Several women were active VWSC members; however, this did not translate into their taking decisions on the design of water delivery structures and systems because decisions for these were predetermined in the project guidelines. The public stand posts consisting of a tap stand and a platform with no cover continued to be built, even though they lacked privacy for bathing, washing private clothes, and so forth. They were inappropriate for these tasks, resulting in women making several visits to multiple water sources to meet different household and private needs. Women still do not touch these stand pipes when menstruating, yet the notion of women's impurity in relation to accessing water has never been discussed.

Gender-role reversals were difficult to put into practice. In Himtal village, with an all-women's committee, construction of technical systems remained a masculine responsibility. Men were paid for construction work, while women did the 'voluntary tasks'. Members of the highly praised all-women committee said:

> We participated because the men refused to do all this voluntary work for no gains. It has been an enormous struggle to complete the project and to meet our various work demands at home and in the field. We are unable to change anything in the project design. We do not get any monetary compensation, not even to meet travel costs to the NGO and Project Management Offices. We think this is the price we poor women must pay for getting water.

Participation in the VWSC or the numerous other committees, even for the more vocal and powerful women, has brought few changes to the dominant patriarchal social order. When asked what changes had come about through this participation, women identified with zeal their ability now to challenge the state, represented by local and regional officials and agencies. Yet, these same women shied away from and almost vociferously defended the inequality in household roles, responsibilities and resources between women and men, and across rich and poor households.

The high profile of the SWAJAL project in Himtal owes much to its former president, Hema Devi, who relentlessly (and voluntarily) participated to achieve project demands. This included making a trip to distant Delhi for the purchase of cement and pipes: 'I spent a terrifying night, when they stopped the truck on a lonely road and all the men got drunk.' Though seen as an example of empowerment, Hema Devi found that, in fact, little had changed at home or in the local social order. Mid-way through the project, Hema Devi's young son (of Kshatriya caste) eloped with a married Brahmin woman of the same village. The dominant Brahmin community of Himtal did not tolerate this well; they blamed Hema Devi for the situation, which they saw as a challenge to the social fabric of caste and gender hierarchy. Little blame was placed on the Brahmin family involved, but Hema Devi's family was ostracised for several months and she was asked to resign from managing the water project. When she tried to resist, the village leadership entangled her family in legal complications and she was subject to enormous social and political harassment. No one assisted her, not even the NGO staff who had promoted her 'empowerment'. The other women committee colleagues expressed sympathy but could not reverse the actions of the elected village head, a powerful Brahmin male. Many of the allegations had subsided a year after the event, but she still cried when she talked about her experiences.

Income Generation and Empowerment

An analysis of the sewing training, the SWAJAL project's major income generation programme for women, carried out in Mala village showed the programme had been of use only to richer women who had sewing machines to practise on. There was, however, no evidence to show that they had made an income from sewing or that the programme had brought about increased gender equality. The poorer women without sewing machines, the majority, had forgotten what they had learnt six months later. The bitterness of these women in realizing their inequalities was evident: 'we sewed and saw our differences, but we could do little to change this'.

Local feminists note that

women in the hills have long been used as tools and instruments to lure development funds by both NGOs and the government. Yet projects focussing on women have rarely changed the lives of the women. The bending of women over project funded sewing machines and the vacant looks of women knitting blankets and *dhurries* tell a story of despair. The answer lies in liberating women within their homes and enabling the breaking of socially bound shackles to liberate their conscience. (Upreti and Pant 1991:1)

The Case of Bina Devi

Bina Devi, a single parent with two young children whose husband, working in Mumbai, pays only occasional visits, heads the lone Dalit household in Mala. Socially and geographically isolated – as a result of social segregation – and among the poorest households within the community, she genuinely qualified for every one of the project's benefits; yet, not only was her household excluded from all the project had to offer, they took away her only water source. Her small terraces of rain-dependent agricultural land were traditionally fed by the family's informally owned water source, the spring that supplies Mala with water for the SWAJAL project.

> See, that area is mine; it is where I collect firewood and fodder. All women here in the village 'own' such pieces of land, bordering the government-owned forests. This is the only thing we women actually have ownership of. The water source was located there and this was what I used. When the project was beginning the Committee members came to tell my mother-in-law, who was here then, that they were using the source for the project, but that we would be given access from the pipes. We were not asked, we were told, and the fact that we could not stop this perhaps points out our position in the village. Any other household would have objected to this plan, if the water was in their land. My mother-in-law told them that she did not believe them, but I was still hopeful. I so wanted to connect to others in the village; I feel so very lonely and alone. But the promises, as always, did not hold. I now 'share' water from another source. (Bina Devi)

As a Dalit, Bina Devi lacks social access to the nearest source of clean water, a stand post three kilometres down a steep slope from her house, which is used by her upper-caste neighbours. When she demanded a stand post near her house, she was initially told she must pay for an individual connection, although her isolation was not by choice but by social compulsion. After consultation with her husband, she agreed to pay charges. Since all this had taken a year, she was told that she must now pay double the charges and also buy the necessary pipes. On her husband's next visit, they bought the pipes and carried them four kilometres from the road head, but were told that a connection could not be provided; various theories are quoted locally to explain why the family could not be included.

Bina Devi's water source is now a cement tank fed by diverted *gad* water and shared with a Brahman family, who live in a nearby hamlet. When, during the monsoon, water flows through an open tap halfway up the tank, she can use it, as long as she touches neither the tank nor the tap. When levels decrease in summer, water needs to be scooped out of the tank, and Bina Devi must wait for the Brahman woman to give her water; neither she nor her young children can access this water on their own. During summer there is often not enough water for two families, and because Bina Devi always gets what is left over, she needs to walk almost every day down to the river located a kilometre from her house, across a steep hill, an enormous demand on her energy and time.

Despite the obvious norms for Bina Devi's mandatory inclusion in the VWSC, she was not represented there. She was not called for any project meetings and was unaware of what was happening. Upper-caste women VWSC members claimed either, 'we called her but she did not come,' or, 'the house is not within the village boundary'. The (male) committee president disputed this: 'it is in the village, but she is a single woman and she does not want to/cannot attend meetings'. The strong and enthusiastic treasurer, who worked diligently for early project completion, was among the most vocal in declaring that Bina Devi should not have access to any project benefits. In the local reality, Bina Devi's touch and mere presence were considered polluting. NGO field staff admitted candidly, 'to insist on the Dalit woman's representation would have antagonized the dominant higher-caste community in the village. This would hamper completing the project on time, which was our major responsibility.' Senior NGO staff said they were unaware of Bina Devi's situation.

Bina Devi felt powerless: 'Who do I ask? Mine is the lone scheduled caste family in this village. Who will listen to me? We have no money to do anything ourselves. My husband's salary is so low that he can barely manage to send money or even come home.'

Gender and Caste within the NGO

The disparity observed at the household and community level is reinforced in the structure, culture and functioning of institutions. The NGO's commitment to community empowerment is widely respected, but its vision is narrow and while the organization works primarily with women, the battle is restricted to women's struggle with the official bureaucracy and against alcoholism amongst men; more sensitive issues of social inequality at community level are rarely challenged. Like many NGOs it upholds the sanctity of culture (Mukhopadhyay 1995) and defines the more sensitive issues of gender (and caste) inequality as a brand of Western feminism 'which will not work here'.

The NGO founder members, who are also the senior management staff, are all Brahmins; amongst over a hundred staff, there are only two Dalits in lowly positions. The organizational commitment to community empowerment is not reflected in a commitment to overcome caste-based inequities internally or in the hill communities.

There is little consideration of gender issues within the organization, apart from vague policy directives to recruit women; even these face resistance. The view of one male founder member about increasing women's involvement in senior management was relatively mild: 'we would like to gender balance our team, but we will not do this just to please donor demands of gender balance'. There are, however, many more women staff at the field level and perhaps the next generation of senior staff will be more gender balanced.

The organization values the 'voluntarism' of community development, yet the brunt of this is borne by the mainly female field staff, who work hard for very basic salaries. Working conditions are difficult and turnover is high especially amongst women, most of whom are unmarried or have chosen to remain single. One woman said, 'I have worked here for nearly three years now, and literally every day. I will certainly not be able to continue with this when I marry.' Field staff endeavour to meet project and programme demands but there are no personnel rules, formal or informal, to guarantee fair employment terms, and little consideration of gender-sensitive employee needs. The advocates of equity and empowerment often ignore these issues in their own backyard.

Hira lives in the area where she works. Her disillusion with the organization, which she adopted as her second home, is clear:

> I worked twenty long years for this NGO, travelling all day on foot across difficult mountain terrains. More than that I defied the abuse heaped on my morality, when I left home to work outside. But what do I have to show for it now? It has been a few months since I became completely ignored and forced to stop working because I did not concede to the director's wife's demands. I had imagined that I could be strong and resistant, which was what I encouraged women in the village to be. I did not realize there was a difference between what we can say and what we can do. I am poorly educated, unable to speak and write English, and have realized in these past few months that my years of experience are of no value in the market. I am the only employed member of my family but am now back to where I started.

Conclusion

Based on this case study it is clear that gender theory captures the field realities in these villages, where gender, caste and wealth shape communities and result in the unequal allocation of roles and responsibilities, and unequal access to and control over key resources – including water. While existing water policies and projects highlight the need for women's empowerment and stress the essential role they play in the provision of clean drinking water for all, they still fight shy of tackling the complex social relations and institutions that uphold these inequalities.

Rather than address gender inequality and women's subordination, or the exclusion of Dalits, the programme discussed above tried to promote women and Dalits in the villages without providing the wide-ranging support or the fundamental shifts needed to ensure their position is changed and that they can meet new challenges and play new roles. By ignoring the complexities, and opting for simpler approaches of, for example, offering committee places, training and access to income generation to vulnerable and disadvantaged groups, the project ensured the replication and perpetuation of the social relations of inequality. Consequently, some of the most needy and vulnerable failed to benefit from the new supplies of water that ostensibly were designed primarily for them.

While women's participation in water projects holds the promise of being meaningful it can only be so if it involves awareness about, and commitment to, reducing the inequality of socially allocated roles and responsibilities for water, and inequity in water access. Further, women, especially those disadvantaged by caste and class, must be enabled to really influence the planning, design and management of water delivery systems in ways that are appropriate, adequate and reliable for them. This requires a good analysis of these issues by the project management and proper allocation of resources to achieve these goals. Addressing gender and exclusion issues involves targeting unequal social relations, and the beliefs and cultures that structure and justify them, yet all too often gender is misunderstood and practised simply as 'women's increased participation' in water programmes and the most vulnerable in communities are left behind.

Acknowledgements

Acknowledgement is due to both Tina Wallace, who provided expert knowledge and support to the author in the writing of this paper, and Mary Lloyd who greatly helped the author with editing parts of her PhD into this chapter. Their hard work was invaluable.

References

Agarwal, B. (1992), 'The Gender and Environment Debate: Lessons from India', *Feminist Studies*, 18(1).
—— (1997), 'Environmental Action, Gender Equity and Women's Participation', *Development and Change*, 28.
Baden, S and A.M.Goetz (1998), 'Who needs Sex When You Can Have Gender? Conflicting Discourses on Gender at Beijing', in C. Jackson and R. Pearson, eds, *Feminist Visions of Development: Gender Analysis and Policy*, London: Routledge.
Boserup, E. (1970), *Women's Role in Economic Development*, New York: St. Martin's Press.
Chibuye, G. (1996), 'Gender Perspectives of the Water Sector', *Water Sector News*, 4.
Cleaver, F. (1997), 'Gendered Incentives and Informal Institutions: Women, Men and the Management of Water', in D. Merrey and S. Bhaviskar, *Gendered Analysis and Reform of Irrigation Management: Concepts, Cases and Gaps in Knowledge*, Sri Lanka: International Water Management Institute.
Cleaver, F. (2001), 'Institutions, Agency and the Limitations of Participatory Approaches to Development', in Cooke and U. Kothari, eds, *Participation: The New Tyranny*, New York: Zed Books.
Cleaver, F. and D. Elson (1995), 'Women and Water Resources: Continued Marginalisation and New Policies', *Gatekeeper Series*, 49, London: International Institute for Environment and Development.

Davidson, J. (1990), 'Restoring Women's Links with Nature', *Earthwatch*, 37, Oxford.

Dayal, R., C. Wijk and N. Mukherjee (2000), *Methodology for Participatory Assessments with Communities, Institutions and Policy Makers: Linking Sustainability with Demand, Gender and Poverty*, The World Bank, Washington: Thomson Press.

Dixon, R.B. (1985), 'Seeing the Invisible Women Farmers in Africa: Improving Research and Data Collection Methods', in J. Monson and M. Kalb, eds, *Women as Food Producers in Developing Countries*, Los Angeles: University of California Press.

Elson, D. (1991), 'Male Bias in Macro-Economics: The Case of Structural Adjustment', in D. Elson, *Male Bias in the Development Process*, Manchester: Manchester University Press.

—— (2003), *Guidelines on Swajaldhara – Swajaldhara Gram Hamara*, Rajiv Gandhi National Drinking Water Mission, Department of Drinking Water Supply, New Delhi, Ministry of Rural Development.

Guijt, I. (1993), 'Water and Gender in Agenda 21', Paper presented at the conference on Gender and Water Resources Management: Lessons Learnt and Strategies for the Future, Stockholm, SIDA Workshop, December 1993.

Harrison, P. (1993), *Inside the Third World: The Classic Account of Poverty in the Developing Countries*, London: Penguin.

INSTRAW (1989), *Women, Water Supply and Sanitation: Making the Link Stronger*, Dominican Republic: INSTRAW.

Jackson, C. (1993), 'Doing What Comes Naturally? Women and Environment in Development', *World Development*, 21(12).

Jacquette, J (1982), 'Women and Modernisation Theory: A Decade of Feminist Criticism', *World Politics*, 34(2).

Joshi, D. (2004), *Whither Poverty? Livelihoods in the Demand Responsive Approach: A Case Study of the Water Supply Programme in India*, London: Overseas Development Institute.

Kabeer, N. (1994), *Reversed Realities: Gender Hierarchies in Development Thought*, London: Verso.

Kamminga, E. (1991), 'Economic Benefits from Improved Rural Water Supply: A Review with a focus on Women', *IRC Occasional Paper*, 17, The Hague: IRC International Water and Sanitation Centre.

Leach, M. (1991), *Gender and the Environment: Traps and Opportunities*, Sussex: Institute of Development Studies.

Leach, M. and C. Green (1995), 'Gender and Environmental History: Moving Beyond the Narratives of the Past in the Contemporary Women–Environment Policy Debates', *IDS Working Paper*, 16, Sussex.

Mehta, L. (2000), 'Water for the Twenty-First Century: Challenges and Misconceptions', *IDS Working paper*, 111, Sussex.

Mies, M. and V. Shiva (1993), *Ecofeminism*, London: Zed Books.

Mohanty, C. T. (1991), 'Under Western Eyes: Feminist Scholarship and Colonial Discourses', in C. Mohanty, A. Russo and L. Torres, eds, *Third World Women and the Politics of Feminism*, Bloomingdale: Indiana University Press.

Molyneux, M. (2001), *Women's Movements in International Perspective: Latin America and Beyond*, New York: Palgrave.

Moser, C. (1993), *Half the World, Half a Chance. An Introduction to Gender and Development*, Oxford: Oxfam.

Mukhopadhyay, M. (1995), 'Gender Relations, Development and Culture', *Gender and Development*, 3(1): Oxfam.

Oakley, A. (1972), *Sex, Gender and Society*, London: Maurice Temple Smith Ltd.

Prokopy, L. (2004), *The Relationship between Participation and Project Success: Evidence from Rural Water Supply Projects in India*, West Lafayette: Purdue.

Rodda, A. (1994), *Women and the Environment*, London: Zed Books.

Rogers, B. (1980), *The Domestication of Women: Discrimination in Developing Societies*, London: Kogan Page.

Sara, J. and T. Katz (1998), 'Making Rural Water Supply Sustainable: Report on the Impact of Project Rules', Washington, DC: The World Bank.

Schouten T. and P. Moriarty (2003), *Community Water, Community Management: From System to Service in Rural Areas*, London: ITDG Publishing.

Shiva, V. (1989), *Staying Alive, Women, Ecology and Development*, London: Zed Books.

Upreti and Pant (1991), *Uttara*, Nainital, Uttaranchal.

Visscher, J.T. (2000), 'Concept Proposal for Operationalising Gender in Integrated Water Resources Management', Presentation at the Water and Gender Session at the Second World Water Forum, The Hague, 17 March 2000.

Whitehead, A. (1979), 'Some Preliminary Notes on the Subordination of Women', *IDS Bulletin*, 10(3), Sussex.

World Bank (1996), 'Rural Water Supply Project to Bring Clean Water, Sanitation Services to 1,000 Villages in India', *World Bank Press Release*, 96/30/EAP, http://www.worldbank.org/html/extdr/extme/9644sa.htm.

—— (1998), India – Water Resources Management: Sector Review, Rural Water Supply and Sanitation Report, in cooperation with The Rajiv Gandhi National Drinking Water Mission, Ministry of Rural Areas and Employment, Government of India.

9

Enabling Women to Participate in African Smallholder Irrigation Development and Design[1]

Felicity Chancellor

Introduction

It is widely recognized that women play a prominent part in crop production in Africa. They are particularly active in the smallholder irrigation sector but take almost no part in its design processes and have extremely limited roles in management, resource control and strategic decisions. Male irrigators too suffer limited access to decision-making processes but, because women fulfil such a prominent role in cultivation, and because their constraints are more severe, it is now necessary to consider explicitly their needs to ensure that smallholder irrigation fulfils its potential in contributing to sustainable rural livelihoods.

Although smallholder irrigation is rightly perceived to have the potential to reduce poverty, many existing schemes in southern Africa perform poorly. The study described here explores constraints to successful irrigation in southern Africa and identifies gender disparities associated with particular aspects of design as a major contributing factor. Lack of user participation at the outset of projects is closely associated with poor performance and even eventual abandonment of irrigation. Specific tasks such as land preparation, maintenance and marketing, illustrate disparity between needs and opportunities for men and women irrigators. Recognizing that advocacy for women's rights is a wide and difficult issue, this discussion is limited to highlighting the need for design to be informed by understanding of the social division of labour and the networks that support the rural resource base.

Sound investment in people to enable and empower them to improve their livelihoods demands gender-aware approaches that influence the earliest stage of the development process. If the potential of irrigated production is to be fulfilled, whilst keeping quality of life in focus, both men and women must contribute effectively to design and continued development of user-friendly small-scale irrigation systems (Boelens and Davila 1998). There are unlikely to be quick fixes. Thus education and training are the keys to enable rural people to develop confidence in their ability to work with designers and participate in decision-making in a way that will enable designers to take account of the full range of user needs and limitations.

The Need for Special Attention for Women

Investment in smallholders has been identified at the World Bank as *the* investment that will contribute most to development in sub-Saharan Africa (Bautista et al.1998) and it is widely recognized that in smallholder irrigated farms women play a prominent part in crop production (Chancellor 1997). Hitherto women have had limited opportunities to take part in the decision to adopt irrigation or to define the form it might take. The inconvenient work schedules and difficult conditions of work that result limit returns to investment. Furthermore, social and cultural factors limit women's access to productive resources, key institutions and technical and agricultural services. Thus women's work in the smallholder irrigation sector is often characterized by poor productivity, fragile sustainability and insufficient reward. The sector is strongly influenced by the fact that women, with these attendant problems, play a major role in providing labour. Yet in the twenty-first century many African countries will look to smallholder irrigation to provide the much-needed increase in food to support rapidly growing populations. The time is therefore ripe to analyse carefully the part gender issues play in constraining the success of smallholder irrigation and work towards better practices.

The Gender Sensitive Irrigation Design Study (GSID) in southern Africa was commissioned by the Department for International Development, UK, to explore constraints to successful irrigation. This research responded to the perception that women irrigators faced a greater struggle than men to produce crops and profit from farming. It sought to identify strategies that might redress this imbalance through studies conducted in Zambia, Zimbabwe and South Africa over a two-year period in which fourteen irrigation schemes were investigated. The schemes were all small-scale, year-round-irrigation systems, with plots of around one hectare or less, chosen to be representative of the smallholder irrigation sector in southern Africa, some relying on gravity flows and others on pumps and sprinklers. A range of research methods was used, beginning with an initial survey and inductive analysis followed by focus groups, key informant interviews and participatory ranking exercises. The study showed that some constraints were indeed brought about by lack of attention to gender issues in design of irrigation projects and by the lack of people-centred design. It investigated how men and women are bypassed in decision-making and identified gender disparities associated with particular aspects of design. Poor and insufficient participation at the outset of projects was often to blame for poor performance and institutional arrangements were also a major source of dissatisfaction. The connection between irrigation design and workloads is poorly understood. Farmers, particularly those who have no irrigation experience, do not readily appreciate the full implications of adopting irrigated production. Designers are preoccupied by efficient water distribution and may have limited experience of farming tasks (Chancellor and Hide 1997). Thus participation, unless it is preceded by awareness raising and meaningful discussion between farmers and designers, does little to minimize work. The dynamics of working relationships between men and women and the various essential tasks they do have to be understood in order to choose the most suitable technology. This necessitates raising the profile of participation, making it a process that contributes to success rather than one where only vague approval is sought. Mainstreaming gender in this arena requires substantial shifts in attitude among designers, planners and communities themselves before effective gender sensitive strategies and designs can emerge.

General Issues in Smallholder Irrigation

Sustainable Livelihoods

When people take up irrigation they start an intensification process: each irrigated acre needs more preparation, more investment of infrastructure, more water, more input and more labour than before to ensure significantly greater benefit. This intensification can only be sustained if produce is highly valued either as a direct contribution to livelihoods or as a source of profit. Motivation must exist to meet the high costs (Ashley and Carney 1999). Unreliable irrigation water supply causes disaster, only more intensely than the failure of rain on an extensively grown crop. Serious interruption to the supply of irrigation water will reduce or eliminate returns to all the extra effort. When inputs are obtained on credit, losses that result from water shortage attract interest and rapidly cripple poor farmers. Thus, the increase in potential so often advertised in promoting irrigation comes hand in hand with increased risks and less-advertised vulnerability, yet safeguards are commonly neglected.

Even without the added burden of credit, regular payments such as electricity charges and water bills must still be met. Many of the poorest farmers, including elderly people, widows and single mothers, and those who care for relatives or orphans, are particularly vulnerable to losses and often barely cover their costs from year to year. This is frequently the result of factors beyond the control of the cultivator, like system malfunction, default on the part of a supplier or service provider, or contractual and marketing problems. The positive impact of irrigation on sustainable livelihoods is reduced by these cases. In making irrigation user-friendly for these people, enormous gains can be made in the fight against poverty.

A study by Merle in Limpopo Province, South Africa revealed that only a small percentage of smallholder irrigators were able to make a reliable profit from irrigated agriculture (Merle et al. 2000).

Land Tenure

Ownership of irrigated land is often seen as a key to encouraging people to invest in long-term strategies to preserve water quality and supply and soil productivity and to establish sustainable business links. Tenure arrangements often explain farmers' attitudes to undertaking long-term investment such as major maintenance or introducing tree crops. In Africa, relatively few smallholder farmers hold title to land and very few of those who do are women. This lack of secure tenure reduces motivation to plan for the long term. In some areas land rights are considered secure by farmers but the lack of title is still a factor in reducing the availability of production credit. In Kenya and in Egypt, where ownership is more common, investment and crop diversity flourish and a market in renting irrigated land has developed to benefit tenants and landowners. Where land ownership is predominantly communal, traditional leaders confer permissions to occupy and use land according to the local custom. This provides relatively secure tenure, and rules are well established, though generally within a patriarchal framework that significantly favours men. Nonetheless women acting in groups are nowadays successfully gaining permission to use land for irrigation and credit for production. The international NGO CARE is particularly noted for promoting women's irrigated vegetable growing, often setting up gardens that provide large numbers of women with small plots to supplement diet and income using simple irrigation methods. This model is being taken up by other NGOs and by in-country programmes in the region. In Eritrea, women's groups sometimes apply to local elders for distributions of land for irrigated vegetables and are successful in gaining favourable conditions of tenure.

Intensification requires labour and provides potential to improve livelihoods for landless people. Poor women figure prominently among the labourers. However, they tend to suffer first if the system fails. Overall, local custom has not kept pace with demographic and social change brought about by out migration and the HIV pandemic, so that the greatly increased numbers of woman- and child-headed households are not well served by established custom. Where new irrigation schemes are developed, there is an opportunity for the rules of tenure to redress these imbalances. However, in most cases inequitable tenure is likely to feature for some time to come.

Raising Awareness

Farmers need good accurate information to consider irrigation. As well as the natural tendency of irrigation agencies and donors to emphasize potential rather than cost, their ability to estimate cost is limited by lack of practical experience in monitoring. Thus the danger of luring farmers into unsustainable investment is real and could increase poverty. Although increased poverty was rarely found in the GSID study, except where pump failure was a feature, it was common to find minimal and inequitably distributed increase in income. Among women, the extra labour involved in irrigation increased their relative poverty, unless the proceeds from the irrigated crop were shared fairly. In Zimbabwe, Matshalaga reported that women complained that they weeded conscientiously but saw none of the profit when the crop was sold (Chancellor et al. 1999: Part 6). Long hours and year-round cultivation draw women's energy away from other family livelihood activities such as basket and beer making. However, because household or farm incomes rise, women's relative poverty goes unnoticed. Greater awareness of the reality of intra-household income distribution will assist designers to analyse user motivation for particular irrigation practices. For example, reluctance to weed is often seen by the outsider as a manifestation of ignorance or intransigence but there may be sound reasons behind this behaviour from women's viewpoint.

Where demand-led projects are promoted, women and poor people are often unenthusiastic, despite the intention that exactly these people should be the beneficiaries of small-scale, low-cost irrigation. Poor communication of the full range of irrigation options and the many possible operation and management arrangements may explain this apparent lack of demand. To date little attention has been given to providing accurate and clear information flows about technology to semi-literate rural people. Translation services are often missing or of poor quality and little attempt is made to relate technical messages to familiar concepts and available equipment.

In a 2003 study, information flows to women irrigators were found to be inferior to those to men irrigators, most women at best receiving second-hand and inaccurate information from male relatives (Chancellor et al. 2003). Findings from South Africa, Zimbabwe and Zambia highlight the relationships between design characteristics of irrigation programmes and projects, and the workloads and managerial responsibilities of men and women irrigators. However, these issues were seldom discussed with farmers in the initial stages of irrigation development planning. The study sought to identify the impact of gender issues by taking account of project environments, local institutional and cultural frameworks, support provided by agencies and commercial organizations, and attitudes of men and women to managing irrigated agriculture.

Understanding 'Who Does What' in Irrigated Agriculture

Men's Views and Women's Views

When men or women are asked about labour contribution to irrigation, very different descriptions emerge. Divergence of male and female views is as prevalent among African smallholder households as anywhere in the world. In Zimbabwe, both men and women regard themselves as *the* major players in irrigation work, although observations tend to support women's claims. Assessing workloads through participatory processes alone appears to be inadequate; there is also a need to understand people's priorities relating to the various elements of the workload. Additionally, large parts of an irrigator's workload may be outside the irrigation sector, necessitating a broad approach to include all relevant factors. The inescapable load of domestic provision falls to women and women in addition often work as labourers on other plots. Men too have additional workloads, some in formal employment that they try to conceal for fear of losing their right to cultivate, others in agricultural enterprises, businesses, religious duties or traditional authorities. There is a clear need for broader and deeper analysis of workloads in a livelihood context.

Assessing how individual workloads are affected by design requires participation. Many conditions combine to make women's participation especially difficult and to constrain dialogue with women in irrigation development. Issues of land tenure, water rights, education, technological know-how and confidence, all appear complex and reduce the propensity for women to participate and for agencies to explicitly target them; to these can be added social pressures, over-committed time, and low motivation, to complete a discouraging picture. Poor gender awareness among irrigation professionals and institutions exacerbates the situation. Despite these factors some women participate to good effect, particularly in small vegetable growing developments, often alone but commonly as a result of NGO activity specifically targeting women.

Up to 40 per cent of households in some areas operate without regular adult male help. Nonetheless, complaint relating to workloads is unusual, because hard work is valued in local culture and arduous tasks are accepted. The same values lead to minimal attention to the labour saving characteristics of the tools. Women hardly consider trying out long-handled hoes for fear of being labelled lazy. Social taboos also explain women's reticence in contradicting or questioning the views of male relatives in public. Survey techniques must take such parameters into account to encourage women to be forthcoming about work. Survey analysis leads on to a focus-group approach, using various groupings of people such young or old men, widows or unmarried mothers, old people, young unmarried males and so forth. It is a mistake to assume that sex is always the most important axis of differentiation but interviewing groups of men and women separately at least some of the time brings out some very interesting insights and reveals how important male and female cultures are in moulding people's perceptions. Age, wealth, education, clan associations, and life cycle are important determinants of social behaviour within the broader context of learned gender roles.

Clearly some tasks were recognized as particularly hard work and it rapidly transpired that men and women had different views about the degree of hard work involved. In addition to a strong culture of gender-based allocation of work, the importance of the dynamic relationship between men and women and between tasks became apparent. People's complex work patterns are based on interwoven activities. Thus one person's workload often depends on how well others have done their tasks. Women's weeding is easier if deep ploughing has been achieved by the men. Women therefore have an interest in helping with the ploughing. High work intensity is often made more acceptable if it is interwoven with social activities such as the information gathering and story swapping that come with the drudgery of carting domestic water or watering crops by bucket. Also, tasks incur vastly different workloads for different people. For example, ploughing half an acre is relatively easy for a man who has a team of oxen and a plough, and very hard work for a woman with a hoe.

Additionally, although female-headed households have difficulty mustering sufficient energy, it is also common but less recognized that they experience difficulty with financial and social aspects of hiring assistance. Married women are known to forbid their husbands to plough for widows, lest they be 'loose'. Such complexities have implications for professionals in the sensitive handling of discussion in the participatory process.

Striving for Balanced, Gender-sensitive Views

The contrast between the views and comments of men and women irrigators suggests that participatory approaches must be more focused than in the past. Using the gender, poverty and age axes in relation to discussion of workload and intra-household relations might provide greater insight for designers and encourage responsive designs.

Further contrast was noted between the views of departments and agencies on past participation and those of farmers. Farmers, and more particularly women, often felt their views were unheard, misunderstood or disregarded or that the necessary compromise was unfavourable, whereas designers and developers felt that they had listened to farmers and yet obtained poor co-operation in return. The sources and causes of these mismatches deserve attention.

Professionals in southern Africa were generally disappointed by the performance of the participatory process and by the performance of the completed irrigation schemes. They identified lack of commitment on the part of smallholders, inadequate finance and poor policy environment as contributing causes. These professionals feel frustrated. On the other hand, although irrigators agree that irrigation has brought profit and livelihood improvement they are aware that the schemes for the most part perform below potential. A high level of frustration exists among them too and this is particularly evident among women. Although women's frustration arises from inappropriate institutional arrangements and lack of information, their greatest grievances are about lack of resource control, and relate to the low reward for their labour, which is provided at high personal cost.

These last they identify as a community and cultural issue. Professionals are often unaware of the extent to which community and cultural values mould the ways in which people use technologies, which they regard as simple and straightforward. This is not simply a case of foreign donors' failure to understand the culture in developing countries, but a more fundamental difficulty about learning to listen and understand the context in which the user operates; fitting the hardware to the people not the people to the hardware.

Focus group discussion revealed initially that women value irrigation highly because of their responsibility for family food security. However, further discussion suggested that the major role women play at field level is often not a matter of choice and does not necessarily contribute to family food security (Matshalaga 1998). Similar phenomenon have been recognized since Chambers described inequity at Mwea in Kenya in the 1960s, and one case was well described by Dey and others among farmers in the Gambia in the 1980s (Dey 1990, Carney 1988). Married women are coerced to contribute long hours of labour in irrigated fields but are poorly rewarded in relation to the work done. Where the crop is a 'cash' crop, such as tobacco or tomato grown under contract, the payment often goes directly to the male in whose name the land is held and cannot be accessed by the women. Women are often forced to reduce the hours they devote to their own small money-making schemes such as vegetable growing and craftwork or to skimp on childcare and cooking. Widows considered themselves in some sense well off because they are able to manage their limited resources autonomously. However, young widows suffered the additional social difficulty of being regarded as potential husband stealers. Widows whose male relatives took a managerial role were also less able to profit from irrigation. Identifying different subgroups and talking separately with them was essential to this level of understanding.

An important first step seems to be to identify the types of household that are, or will be, involved in irrigation and to tailor participation to allow their concerns to be aired freely. Identification of the relevant subgroups and avoidance of assumptions that treat men and women as homogeneous groups are recommended. Among men for example there are significant differences between the needs of youths and those of older men. While the young man tends to have energy, ideas and strength, he often lacks access to land and capital. Meanwhile the old man may be unable to use land and productive assets to the full because he is tired.

Before discussion can be helpful, dissemination of information has to take place so that individuals and communities can assess the gaps in their knowledge and marshal their contributions and queries. Information must be well thought out and clearly presented with plenty of opportunity for questions and answers and development of alternative ideas. Gender sensitivity is important in communicating information; in Zimbabwe, for example, it was found that while football analogies were excellent for male irrigators they did not help inform the women.

A balanced view, however, does not necessarily mean one to which exactly the same numbers of men and women contribute. A much more dynamic approach is needed. A balanced view of component parts depends on consideration of where responsibility for each specific task lies. Ideally a balanced approach should take links to other tasks into account and be aware of the irrigator's objectives. Consider the example of land preparation, which, traditionally done by men, is now increasingly left to women. The impact of land preparation on weed growth and water penetration is far reaching in terms of resource-use, workloads, water-use and yield. Strategies to improve the way men and women contribute to land preparation depend to a great extent on improved understanding of the relationship between the various cultivation tasks and on women's empowerment in intra-household bargaining. Design of irrigation systems dictates the land preparation task, long-furrows demanding different skills and equipment to basin cultivation. This too has implications for who will undertake the task within the family.

Mass meetings with voluntary attendance are not an effective way to approach information gathering or a good way to introduce new ideas or develop new strategies. In southern Africa women are reluctant to argue with male relatives in such a setting. Discussion in local interest groups is relaxed, more in-depth and less threatening and leads to more helpful exchanges of ideas and information. Just as it is a mistake not to consult women, it is also wrong to assume that only irrigators are concerned with irrigation development. A broad sweep of opinion in local interest groups can reveal important issues that would otherwise go unnoticed by designers and irrigators and subsequently lead to conflict and lack of local support for the irrigation scheme (Chancellor et al. 2003). Subsequently mass meetings may have a role in ratifying plans and allowing public debate. Consensus is important and much is achieved by developing an overall goal to which men and women contribute. 'Win-win' solutions are ideal. They may not always be possible but their incidence can be increased by well-planned participation.

Gender-based Activities, Responsibilities and Benefits

The findings on which this discussion is based can now be grouped according to the key priorities identified for sustainable smallholder irrigation performance. The weight of these several aspects will vary from country to country and place to place.

Participatory and Institutional
Land tenure and institutional arrangements for irrigation schemes often follow the norms of the rural setting. Irrigation design, however, has the potential to lead

change by adopting more equitable terms and conditions. It is therefore important that local stakeholders are seen to participate and are in effect the leaders of change. A number of issues should be taken into account when considering change, such as that male dominance in land control has been entrenched by past allocation of irrigated plots to the male heads of households. The allocation pattern has entitled men to control the benefit of production regardless of the labour contribution of other family members. To relinquish this control is often regarded as sign of weakness and the positive advantages for men and the community at large need to be discussed and recognized if change is to be sustainable. Addressing this type of attitudinal change requires more action than can be generated by irrigation agencies. Allies are needed in local government, education services and the media as well as among the tribal authorities, where they are active.

Male involvement in decisions on water and water management is reinforced by government and agency approaches. Officials, normally men, deal with male landowners, in some cases because agreements reached with women are not legally binding, or talking to women is socially unacceptable. Moreover, men dominate irrigation committees or water-user associations. Higher literacy rates among men, greater male access to leisure time and men and women's perceptions about leadership and use of technology all perpetuate this imbalance. Despite women realising that they are excluded from influence they may be unwilling to combat the situation without support.

Community participation often assists the rich and powerful to appropriate water resources. Women are not generally among the rich and powerful and their relative power can be further reduced by community action. On the other hand the global preoccupation of developers with equality may run counter to the local concept of equity in the community. An interesting discussion on equity and rule-making provided from the South American context by Boelens has relevance in the African context (Boelens and Davida 1998). He calls for support to peasant communities to develop their own institutions through political action that allows people to be part of developing their own rules. Privatization of water is becoming a crucial issue in many southern African countries and this too may have a negative impact on women because of their poverty. Despite this there is a perception among administrators that privatisation provides an ideal opportunity to consider users' needs and give men and women water users client roles. Again, women are disadvantaged in putting their views to those who will decide the privatization issue because they lack registered land or status as household heads, to whom communication is addressed. Gender awareness in the practice of participation and in crafting institutions has the potential to improve irrigation performance and increase co-operative interaction between men and women in the management and allocation of water.

Land Preparation, Cultivation and Production

The study found that women increasingly take responsibility for land preparation and that it can no longer be assumed that men prepare the land. Existing designs therefore often commit women to costly and exhausting tasks because their resources do not match the design assumption (if any existed). Designers of new schemes have the opportunity to consider men and women's capacity for land preparation and how design could accommodate user needs.

Although land preparation is crucial to successful irrigated production and is increasingly the responsibility of women, their opportunity to interact with or influence the designs that determine preparatory tasks is limited. This is due in part to the invisibility of their contribution, in part to the rural cultural environment, and in part to the culture of irrigation and agriculture departments. The issues that need to be addressed are complex and have far-reaching impacts. For example the adoption of long-furrow irrigation demands land preparation that results in a consistent and correctly angled slope in the field. Whereas this can be achieved with a tractor and plough, and to a lesser extent with draught animals, it is very demanding with only a hoe. For women suffering from AIDS the task is impossible. Women's difficulties in hiring men and animals constrain their ability to get land prepared on time. These difficulties generally reduce yield and quality and lead to low returns. The resultant lack of funds for maintenance or water-related charges threatens the sustainability of irrigation systems and can severely erode the livelihoods of all irrigators: men, women and their families.

Another difficulty that faces women in their efforts to prepare land and cultivate their crops is the deterioration in their personal security. The incidence of rape is high and growing, and the targets are increasingly-young girls. Thus women can no longer allow their daughters to work in the fields unaccompanied; nor can they leave them at home. This situation poses a particular problem for women-headed households and significantly reduces the time available for cultivation.

In the past, extension services and other advisory initiatives tended not see women as an appropriate target because women's contribution to irrigation is often invisible, particularly in male-headed households. There is now a perceptible change in the attitude of extension staff that largely recognizes the important role women play in smallholder irrigation. Despite the fact that applying water increases weed growth and considerable emphasis is placed on the need for timely weeding, little attention is given to labour-saving weeding equipment. Weeding is largely women's work, but women's voices are often weak and shouldering hard tasks is perceived as a virtue. Male extension and irrigation staffs ignore this area of potential improvement. Timely weeding determines the success of the crop and is crucial to food security but the hard work involved can damage the health of women and their children. Added to this is the additional responsibility for irrigated agriculture which is falling on women as the demographic character of the rural areas changes. It is well beyond the capacity of most extension services to facilitate debate among farmers as to change in gender-based work allocation principles. Nonetheless recognition of these influences and changes will go some way to encouraging general discussion on the desirability of supportive or co-operative irrigation management systems.

Marketing

Traditionally, women market small quantities locally, while men control larger volumes of cash crop, travel to distant markets and negotiate contracts. Women-headed households have difficulty with large quantities and distant markets; their relative poverty makes hiring vehicles or finding bus fares a significant issue. They also suffer the time constraints imposed by care of children and sick and elderly people, which is an increasing problem as the spread of AIDs advances. New concepts and innovations such as mobile phones, post harvest value-adding, good presentation of produce and new business linkages can help women overcome market restrictions.

Some aspects of women's marketing are worth noting. Married women irrigators lack control of resources and access to proceeds and are poorly motivated to market actively. In contrast de facto women-headed households, single women and widows who make their own decisions and control profits are motivated to improve all aspects. Women working together in a group, often led by the more motivated among them, can successfully improve their situation in relation to marketing and thereby change the behaviour of the male-dominated commercial community towards them. Change in male/female dynamics opens up opportunities to empower irrigators in market activity and improve sustainability by strengthening market linkages. Women find it especially difficult to access small short-term production loans, but they succeed in providing credit on a rotational basis with small savings clubs. Again, single women and widows have more opportunity to exercise skill in allocating cash resources and managing than married women, even if the scale of their operation is very small. The cash income from marketing produce is very important to women irrigators.

Extension services, however, typically concern themselves with assistance to the larger-scale marketing associated with the male farmers, thus neglecting potential for improved service to women, who comprise the majority of irrigators. The existing networks for the sale of produce and supply of inputs are male dominated. Agribusinesses typically employ men particularly for the sale and servicing of machinery and equipment, and rely almost exclusively on male drivers for delivery. Women who can pay for goods and services often find themselves at the end of the queue for attention either because they are women or because they are unfamiliar with the informal procedures (Chancellor et al. 2003)

Design of Equipment, Management and Maintenance

Few irrigators in the study had been party to design except on NGO-assisted schemes and among individual treadle pump adopters in Zambia. In Zimbabwe and South Africa, irrigation development has been part of an imposed settlement strategy; agencies of government often design, manage and maintain irrigation systems but are remote from the users both physically and conceptually. Irrigators often find the managers to be inaccessible and unresponsive. This may in part be caused by the managers' lack of familiarity with the local people. Managers may have an 'engineering' rather than an 'agricultural' background and lack empathy with smallholder farmers. The resultant poor communication sometimes increases the risk of unreliable water supply rather than reducing it, as might be expected.

A major problem with government-run schemes is that the Ministry staff in charge of essential equipment lack incentives to avoid crop failure. Their remuneration is time-based rather than results-based. When failure occurs women, who are short of both money and time, have particular difficulty in contacting distant ministries. Influencing local staff is often difficult too because of social and cultural constraints. All this reduces the ability of women to influence the speed of action to repair, operate and maintain equipment.

In general the equipment needs of both men and women irrigators are poorly understood. Staff caring for machinery on a scheme often lack farm experience and fail to understand the importance of a particular breakdown as far as plant growth is concerned. Women's needs are particularly invisible partly because the collection and dissemination of information is heavily biased in favour of men. Thus, women's complaints, about pump failure for instance, are not heard. Also, the small enterprises that are so important to women's livelihoods are regarded as relatively unimportant.

The failure to pay more than minimal attention to saving women's labour has already been discussed. Simple hoes used for weeding require women to bend continuously but there are social and cultural barriers to adopting labour-saving adaptations which might imply laziness. The mechanization that has taken place eases men's work such as ploughing. Similarly, we know that women's access to mechanized services is difficult. The inclusion of pumps in irrigation design is a major source of risk, high cost and gender-disparity. The roles of men and women in the use of pumps have not been clearly identified and therefore arrangements for operation, repair and financing of pumps are weak. The GSID study revealed a number of instances in which women who relied on pumps were unable to lobby for repair in sufficient time to save their crops. Operation and maintenance of irrigation infrastructure and equipment are poorly organized and the huge potential for female empowerment from developing women's skills to undertake these tasks has been neglected.

These problems become more acute as growth in the number of de facto female-headed households makes it crucial that women are part of the formal network of rights and powers in relation to water and design choices and maintenance regimes for irrigation equipment. If women remain outside the system and do not contribute to policy and planning, they will continue to be hampered by inappropriate infrastructure and equipment; inequity, inefficiency and conflict will persist in smallholder irrigation.

Conclusions

The Future Role of the Small-scale Irrigation Sector

Africa's food gap, as identified by the International Food Policy Research Institute (IFPRI) in their 2020 vision, suggests that rapid increases in production are needed to reduce future demand for food imports to a manageable level. Irrigation departments in east and southern Africa report growing investment in small-scale irrigation developments, much of it in the private sector or resulting from NGO activity (Ministry of Agriculture, Kenya 1996). In these cases management is vested in individuals or small groups committees. On existing, largely state-promoted, smallholder schemes, infrastructure is managed and maintained badly; leaking channels and broken down pumps abound and water distribution is often unreliable and inequitable. Modern governments are concerned to convert these schemes to sustainable projects with poverty alleviation goals.

Schemes must now be commercially viable and farmers must assume the greater part of the cost of providing water for their varied uses (Blank et al. 2002). The greatest threats to viability have been the failures to give women adequate land rights, rights to participation and training, and to consider gender in the operation, management and capacity-building associated with smallholder irrigation development. Addressing these issues in the future will result in improved returns and sustainable irrigation that will have a positive impact on rural poverty.

People-centred Approaches

People-centred approaches are crucial to addressing gender issues in smallholder irrigation successfully and the inclusion of all subgroups is central to this. Gender awareness must not be limited to grass-roots activities or relegated to a 'gender department' but must permeate policy-making and implementation throughout the sector. Successful mainstreaming of gender-sensitive approaches could improve sustainability and livelihood contributions of smallholder irrigation but not in isolation from technical efficiency. It is therefore important that early results of participation provide motivation for continued interest and action to consider the impact of design features on men and women. The marketing of participation is important. Participation should aim not only to reveal the needs of men and women in relation to layout, tenure, operation, maintenance, support services, finance and marketing, but also to facilitate design contributions from participants. Specific attention must be given to ensuring that the results of participation are clearly understood and ratified by all stakeholders, through memos, meetings and contracts.

Gender Roles

Gender-sensitive design refers to design that recognises the different starting points, obligations, constraints and aspirations of groups of men and of women regarding the use of irrigation facilities. A good gender-sensitive design would be one that maximises sustainability and production while empowering both men and women to fulfil their objectives for an acceptable level of effort. Recognizing change is important to effective inclusion of gender considerations but much remains to be understood and done. Timeliness and quality of participation are thus both important issues and depend on the funds and commitment of irrigation and agricultural planners. Unfortunately, participation is often perceived as a one-off activity at the outset of a physical project and is not seen as a skill that requires training. Ongoing, well-managed participation produces good results when it is integrated into the management process as in several Asian countries. Project managers and communities should give attention to developing women's skills, particularly by targeting women with training opportunities in participation, maintenance and management, hitherto virtually neglected. The overall benefit to sustainability and irrigated production that comes from educating women in these areas must be emphasized and illustrated to ensure the support and co-operation of male irrigators. The GSID study found good examples of this type of approach among NGO-assisted projects, where men and women analysed their needs and defined their roles and project aims together with NGO staff. Early projects followed a steep learning curve in identifying critical issues, potential solutions and technical gaps. There were advantages to men in reduced workload and responsibility, just as increased responsibility empowered women, a fact seldom highlighted. Increases in nutrition, income, self-confidence and social capital were evident despite the technical limitations of some of the projects.

Resources are required to raise irrigators' awareness of the gender implications of development strategies, and to engage people equitably, and in depth, in dialogue and choice. Without commitment and funds, participatory decisions are unlikely to be carried effectively through administrative systems to implementing agencies. It is important that the perception of fairness is maintained and that may mean accepting the community view of fairness or at least taking it into account rather than imposing external versions of political correctness and fairness.

It is also relevant to recognize that women have benefited from irrigation both directly and indirectly, albeit not to the same extent as men. Although the cultural environment in Africa is significantly different to that in Asia it is interesting to note that in Bangladesh increased incomes for men from irrigation and moves to mechanization have impacted on women by creating greater scope for non-farm employment. This has lessons for Africa where the issues of landlessness and unemployment in rural areas are increasingly a cause for concern (Angood et al. 2003)

Increased Budgets

Budgets which are provided specifically for awareness-raising and training in design or rehabilitation processes are essential to ensure gender-aware participation. A gender-sensitive approach not only improves performance and sustainability of irrigation development and livelihoods of irrigators but also has a role in empowering women. Lobbying for women's rights is important but, as long as men are not made party to the cause of women, they will continue to consider lobbying as a threat and will continue to oppose it. Therefore emphasis must be placed on including both men and women in the process, on establishing the extent and nature of their responsibilities in the various activities that contribute to sustainable smallholder irrigation and on ensuring strong links between the outcomes of participation and implementation of projects. Much will depend on the financial commitment of donors, departments, agencies and NGOs in the sector, but even more impact can be expected from changes in attitude among irrigation staff and among smallholders themselves. Co-ordination of effort at both policy and practitioner level will therefore be a key factor in success. If gender-sensitive approaches are to significantly improve smallholder irrigation performance, it is crucial that gender does not become synonymous with women. It is therefore always important to deal with the dynamics between the work and responsibilities of men and women and recognize the importance of subgroups created along the axes of age, wealth and cultural norms. The currently fast-changing roles of men and women in society demand constant monitoring and awareness of changing dynamics.

Male Attitudes

Male perception of, and difficulties with, gender policies and approaches must be recognized and taken into account. Male migration, investment in irrigation, and the returns to irrigation are linked in a variety of ways. Very little has been mentioned in relation to the changing role experience of men separated from their parent communities for long periods of time and their need to retain land rights as a symbol of their continued value in the home and village in their absence, and the impact this has on women cultivators when sourcing services and credit need investigation. Nor has this study investigated the role of remitted income in supporting irrigation development, a social issue that has to be recognized and addressed.

As in Asia, women in Africa have largely been 'takers and not makers' of the rules. Also as in Asia, although they are no longer formally barred from the decision-making scene there is still a vast gap in their appearance and activity (Agarwal 1996). Women are under-represented not only in irrigation institutions but also in other local power centres such as local government and traditional authorities. Economic, educational and political empowerment of women is therefore a major factor in improving the performance of smallholder irrigation.

Note

1. This chapter is based on a paper presented at the International Commission for Irrigation and Drainage Conference 'Micro-2000' held in Cape Town, South Africa, in October 2000.

References

Agarwal, B. (1996), *A Field of One's Own: Gender and Land Rights in South Asia*, Cambridge South Asian Studies, Cambridge: Cambridge University Press.

Angood, C., F. Chancellor, J. Morrison and L. Smith (2003), *Contribution of Irrigation to Sustaining Rural Livelihoods: Bangladesh Case Study*, OD/TN 114, Wallingford: HR Wallingford.

Ashley, C. and D. Carney (1999), *Sustainable Livelihoods: Lessons from Early Experience*, London: Department for International Development (DFID).

Bautista, R.M. and C. Thomas (1998), *Agricultural Growth Linkages in Zimbabwe*, IAAE Symposium, Badplaas, South Africa, August 1998.

Blank, H.G., C.M. Mutero and H. Murray-Rust (eds) (2002), *The Changing Face of Irrigation in Kenya: Opportunities for Anticipating Change in Eastern and Southern Africa*, IWMI, Colombo, Sri Lanka.

Boelens R. and G. Davila (eds) (1998), *Searching for Equity: Conceptions of Justice and Equity in Peasant Irrigation*, Assen: Van Gorcum.

Carney, J. (1988), 'Struggles Over Crop Rights and Labour Within Contract Farming Households in Gambian Irrigated Rice Projects', *The Journal of Peasant Studies*, 15(3).

Chancellor, F.M. (1997), *Developing the Skills and Participation of Women Irrigators; Experiences from Smallholder Irrigation in Sub-Saharan Africa*, OD135, Wallingford: HR Wallingford.

Chancellor, F.M. and J.M. Hide (1997), *Smallholder Irrigation – Ways Forward*, OD136, Parts 1 and 2, Wallingford: HR Wallingford.

Chancellor, F.M., D. O'Neill, N.J.Hasnip, J. Ellis-Jones, E. Berejena and N. Matshalaga (1999), *Gender Sensitive Irrigation Design: Guidance for Smallholder Irrigation Development – Parts 1–6*, OD143, Wallingford: HR Wallingford.

> Part 1, *Guidance for Smallholder Irrigation Development* provides guidance distilled from all three country experiences.
>
> Part 2, *Group-Based Irrigation Schemes in Zimbabwe* provides an account of commonly met gender issues.
>
> Part 3, *Gender Considerations Relating to Treadle Pump Adoption: Experiences from Zambia.*
>
> Part 4, *Gender Issues in Smallholder Irrigation Rehabilitation: Cases from South Africa* looks at the challenge of turning round an existing failing scheme.
>
> Part 5, *An Assessment of the Implications of Pump Breakdown and Community Participation in Irrigation Schemes, Masvingo Province, Zimbabwe.*
>
> Part 6, *Consultation on Gender Issues in Smallholder Irrigation*, prepared by Zimbabwean sociologists.

Chancellor, F., M. Upton and D. Shepherd (2003), *Towards Sustainable Smallholder Irrigated Businesses (SIBU)*, OD149, Wallingford: HR Wallingford.

Dey, J. (1990), 'Gender Issues in Irrigation Project Design in Sub-Saharan Africa', in International workshop *Design for Sustainable Farmer-Managed Irrigation Schemes in Sub-Saharan Africa*, February, The Netherlands: Agricultural University of Wageningen.

Matshalaga, N. (1998), *Gender Sensitive Design in Smallholder Irrigation Schemes and Equipment: A Consultancy Report*, IDS, University of Zimbabwe PO Box 880, Harare.

Merle, S., S. Oudet and S. Perret (2000), *Technical and Socio-Economic Circumstances of Family Farming-Systems in Small-Scale Irrigation Schemes of South Africa (Northern Province)*, Synthetic Report, CIRAD Tera, Num. 79/00, November.

Ministry of Agriculture, Kenya (1996) 'Livestock Development and Marketing', in *Regional Workshop on Smallholder Irrigation in Eastern and Southern Africa 24th to 30th November 1996, Nairobi, Kenya.*

10

Water and AIDS: Problems Associated with the Home-based Care of AIDS Patients in a Rural Area of Northern KwaZulu-Natal, South Africa

Anne Hutchings and *Gina Buijs*

Introduction

Water problems in semi-arid Ntandabantu in northern KwaZulu-Natal include lack of access to piped water, poor household water conservation practices and the area's unsuitability for further borehole development. The hilly terrain and dispersed nature of homesteads, linked mainly by rough paths and tracks, effectively isolate the poverty stricken community, which has fallen prey to the current HIV/AIDS pandemic. Important factors influencing the situation are the past high rate of male outmigration to the mines and the present ongoing migration of younger men and women to the towns to seek employment. This leaves to the older women the burden of caring for abandoned or orphaned children and for often-sick adults who have returned to their rural homes. The severe drought of the past few years has further inhibited both the cultivation of food crops and the development of craftwork as a buffer against poverty. This chapter focuses on community understanding and experience of health and home-nursing problems associated with an inadequate and polluted water supply.

Background

This case study is based on fieldwork undertaken betweenm 2000 and 2004 in Ntandabantu as part of ongoing research aimed at developing appropriate home-based care procedures, using indigenous medicinal plants, to address current health problems relating to HIV/AIDS in rural areas of northern KwaZulu-Natal. Participants in a survey, conducted by Anne Hutchings to establish how much was locally known about selected medicinal plants, were also invited to a series of workshops to discuss local health problems. Although a few elderly men attended the meetings, most of the participants were women and many were grandmothers left in charge of homesteads and grandchildren. The participants saw as their main problems the hardships being experienced by the ever-increasing numbers of orphaned children and families affected by HIV/AIDS and poverty. They initiated their own project aimed at alleviating these. This later became known as the Sibusisiwe Orphans and Families living with HIV/AIDS project (SOFAH), which was set up in May 2001 at Ntandabantu with the help of the authors and other members of the University of Zululand to assist those affected by the pandemic. (*Sibusisiwe* is the Zulu name the group gave to Anne Hutchings and means 'we are blessed'.) The local headman gave a piece of land to the group of women who had participated in the workshops for use as a vegetable garden. In 2002 SOFAH became a University of Zululand Foundation project to facilitate access to funding. The project is run by the local women,[1] but with advice and guidance from Anne Hutchings and a team of helpers. They have set up their own gardening, craftwork and health committees. Since 2001 the numbers of sick and dying have increased and the extended drought conditions have wrought further havoc. In 2002 funding from the National Research Foundation's Indigenous Knowledge and Poverty Alleviation programme enabled the authors to undertake limited research into poverty levels and the potential for indigenous knowledge and resource development. Preliminary results from our 2002 survey of 100 of the 400 estimated households in the area indicate that the average household consists of eight members but that 60 per cent of these households have less than R500 available for food per month, indeed often less than R200.

Research Procedures

Research procedures undertaken between 2000 and 2004 include:

- Fourteen community workshops held to discuss health problems, usually attended by about forty participants.
- Eight home visits to families with severely ill or recently deceased members. Follow-up visits were made later. Families were selected according to community perceptions of neediness.
- Four surveys. Two of these used lengthy interview schedules, one on plant usage and one on homestead resources, each of which involved about 100 participants. One shorter survey was conducted specifically to ask questions on water problems and another was addressed to women nursing sick members of the family in their homesteads.

Study Area

Ntandabantu, comprising around 400 homesteads spread over an area of approximately 80 square kilometres, is part of the tribal administrative ward of Gunjaneni, which falls under the recently proclaimed municipal area of uMkhanyakude. It is situated approximately 40 kilometres west of the nearest town, Mtubatuba, and twenty kilometres east of the village of Hlabisa, site of the nearest hospital. Hlabisa district is noted for sending migrant workers to work in the gold mines of the Witwatersrand and Free State. Although relatively close to these centres, Ntandabantu appears to be an almost forgotten pocket. Ntandabantu homesteads were included in the 1998 census conducted by the Africa Centre,[2] but were not included in the more recent health-related research surveys undertaken in the area by the same organization.

Although rural areas of KwaZulu-Natal such as Ntandabantu have names, they are not discernible to the casual visitor as discrete settlements. Like many such places along the eastern coast of South Africa, Ntandabantu is not a village or hamlet comprising a cluster of homesteads but a rural area whose inhabitants live in homesteads strung out along the ridges of hills. These homesteads are separated from one another by distances that range from a hundred metres or so to several kilometres (Sansom 1974). The inhabitants of Ntandabantu speak Zulu, although not all are ethnically Zulu, some having neighboring Swazi or Tonga ancestry. In accordance with Zulu custom, however, the area contains a hierarchy of authority, beginning with the homestead head or paterfamilias. Although he may not be physically present, since many adult men from the area work as migrant labourers in cities such as Johannesburg or Durban, he makes all the major household decisions (Krige 1950, Kuper 1980). Ranked above the homestead head is the local *induna* or headman. The position of headman carries considerable weight in rural Zululand and Ntandabantu is said to be named after the first headman of the area, who was known for his kindness to his subjects, the name meaning 'friend of the people'.

Traditional farming was based on large-scale stock-keeping with a high consumption of dairy products such as sour milk, and with relatively less emphasis on cereals. Much of the income from migrant labour in the twentieth century was reserved for the purchase of cattle. A large herd of cows symbolized a Zulu man's prosperity and status as head of his family. Today, reliance on subsistence farming for a living has greatly decreased and young people hope that their education will be sufficient to provide a job in the towns or cities. Ntandabantu is served by two schools, Ntandabantu CP (primary) school and Ndabeziphezulu High School, each of which is attended by around 300 pupils. The Ntandabantu CP school provided a meeting place for the project's activities and thus started to function as a community centre. Staff and pupils from the High School participated in the various activities held between 2001 and 2003. Until recently these schools had received little support from the provincial government of KwaZulu-Natal but in 2002 the educational authorities made provision for school feeding for four days a week for children attending the primary school.

The homesteads of Ntandabantu are connected by rough and steep gravel paths, which residents have to follow for more than a kilometre before joining the main road. The distance from the main road to the schools is about 3 kilometres and some homesteads are even further away. The hilly terrain and rough and degraded paths are a major obstacle to local people in their quest to obtain water and take sick relatives to a clinic or hospital. Minibus taxis go only as far as the entrance to the main road; most of the homesteads are inaccessible to cars and the sick and elderly have to be transported to the main road in wheelbarrows, which are also used to transport water and shopping. A mobile clinic calls at a local school once a month on pension day[3] if the weather is clement, but stays only for a few hours and often has inadequate medication for patients. There are two municipal clinics on the outskirts of adjacent wards and the nearest of these is 11 kilometres away from their primary school.

Difficulties in Collecting and Using Water

The availability of water in Ntandabantu has always been problematic. Coal seams dominate the geological formation and these have made spring and borehole water brackish and unpalatable. Available sources of water are many and varied but all have severe limitations. There is one river and a few subsidiary streams, which had little water in them when the project started working in the area and have since largely dried up. There is some evidence that the establishment of *Eucalyptus* (blue gum) plantations some distance upstream from Ntandabantu may have contributed to the reduction in the flow of water. It was alleged at a rural appraisal workshop attended by Anne Hutchings in Hlabisa in 1997 that rural people were encouraged to establish these plantations by 'developers' in response to the demand for firewood and sustainable income. Little notice seemed to have been taken of the amount of water these trees consume and the effect on streams and rivers. However, two years later the local *inkosi* (chief) was reported to have banned, on environmental grounds, the establishment of any further plantations in the area (personal communication, Mrs N. Drysdale).

There is a seasonal pond outside the primary school, used for drinking water by humans and animals, but this is frequently dry. Many residents have complained of the pollution caused by the animals. Temporary wells in dried sandy river-beds are dug by the women with occasional help from male members of the family when requested. Water from these wells was used to prepare the school meals, consisting of rice or maize and beans, cooked by the women in three-legged pots just outside the headmaster's study. In 2002 women were obliged to rise at two and three in the morning to carry water on their heads from the nearest hand-dug well, which serves the community on a first come, first served basis. By August 2004 the two local hand-dug wells had dried up and women had to walk for up to three hours to fetch drinking water from a hand-dug well in a neighbouring ward. Residents complain that well water tastes of vegetation but it is still found preferable to borehole water.

Tanker water is supplied by the municipality. In 2002 it appeared to be delivered regularly to the primary school and store, with stopping points for women waiting with containers, but in 2003 delivery became very infrequent and thus tanker water became an unreliable source. A cholera epidemic in 2002 in Zululand meant that local government spent extra on bringing clean water to remote rural areas, but with the reduction in cholera cases in 2003, the extra expense may not have seemed worthwhile.

There was one borehole sunk in 2002 for an agricultural project. Water from this source could be bought and was suitable only for agricultural purposes on account of its poor taste and poor lathering properties. No piping facilities were available and so it was difficult to access for home gardens. In 2003 a type of windmill was provided by the Department of Agriculture next to the borehole and villagers no longer had to pay for diesel fuel for the pump. The problems of the poor taste and the lack of pipes were not solved and by August 2004 even this source had dried up and there was insufficient water to maintain a cabbage growing project.

Very few homes have rainwater tanks. Those that do have frequently not been able to keep them in good repair. Mrs Mbokazi, a middle-aged housewife, said that some time ago, when her husband was employed in Newcastle, an industrial town about 200 kilometres from Ntandabantu, the family was able to afford a local man to build for them two cement rain water tanks, which they still use for drinking water. Mrs Mbokazi said that the tanks fill up in the summer months in the rainy season and she keeps the tanks padlocked to make sure the water is only used for drinking purposes. Other families were less fortunate. The authors observed the remains of a rusted water tank being used as a fowl run in one homestead. Rain water tanks and gutters at the schools have been repaired recently but insufficient rain has fallen to fill them. Teachers, most of whom live outside Ntandabantu, bring their own drinking water and there is not sufficient water to supply hand washing facilities in the newly erected pit latrines.

As in much of Africa, fetching water for household purposes constitutes much of a woman's working day. One participant who works away and goes home for one or two weekends every month, says most of her time is spent fetching water as this is seen as 'women's work'. The extra burden placed on elderly women by absent or ill family members means that children are asked to perform household chores more often than usual. Young boys fetch water but they put plastic drums (usually 25 litres capacity) on a wheelbarrow, three at a time. Fetching water is a daily chore for children, to be undertaken after school. The amount of water fetched is a function of the child's age. A child of ten years would be expected to fetch one drum only in a wheelbarrow; older ones would fetch more, up to three drums.

While it is generally the job of women and girls or small boys to collect water, a man living alone will have to fend for himself. John Dube, a 57-year-old bachelor, said he usually collected water once a day, unless extra was needed for special purposes such as clothes washing at home or irrigating the vegetable garden. It took him about an hour to walk either to the river or to the dam to fetch the water. In the hot summer months this would increase to an hour and a half, including the time taken to fill the plastic drum used as a container. On rare occasions, such as a funeral or wedding, men do fetch water but then they load drums on a trailer drawn by oxen or donkeys. The better off can afford to hire tractors. Extra water is also needed at holiday times (Christmas and Easter) when there are many visitors and additional food has to be cooked and sorghum beer made.

Dube says the community of Ntandabantu would prefer access to piped water or to a borehole for health reasons. He noted that there were too many people sharing the river water and that water was generally scarce. He added that during the recent cholera epidemic (2001–2) 'we used to boil our water. Sometimes we poured Jik (bleach) into the drinking water.' He remarked that drinking water could be contaminated if kept in the kitchen. The taste was affected by smoke from the cooking fire and if a lid was not carefully placed over the water container, cockroaches fell into the water from the thatch of the roof.

Several informants commented on the lengthy preparation needed before clothes could be washed in water collected from wells. This source was preferred to water obtained from the only borehole. One method was to prepare a fire, burn aloe leaves and then collect the ash from the fire. One mug of ash would be added to twenty litres of dirty water collected from the pond. The mixture was left for two hours before washing the clothes. Another method reported involved mixing some cement into the washing water and leaving it to clear for a few hours before washing. Not surprisingly, it was observed that clothes wore out sooner using this mixture and several informants said that it also caused colour to be bleached from clothes. But in August 2004 there was insufficient water to wash clothes and the headmaster of the junior school remarked that the children could no longer be expected to come to school 'clean'.

The Effects of Drought in the Area

Most of the inhabitants of Ntandabantu today attempt to supplement the monthly government pensions for the elderly with some vegetable growing around their homes, but drought has intensified over the years, making it difficult to find water for human or animal consumption, let alone enough to irrigate plants. A regional newspaper reported in May 2003 (*Zululand Observer*, 21 May: 3) that drought continued to plague the province of KwaZulu-Natal 'causing a ripple effect in all spheres of the economy, with increasing poverty in rural communities in many instances leading to crime'. Although Ntandabantu is a rural area, new legislation, which aims to bring all rural areas in South Africa within the ambit of local municipal government, means it now lies also within the jurisdiction of a municipality that attempts to deliver water by tanker to areas with none. A further repercussion mentioned in the article is that with the sole daily focus of the community on collecting water, productivity has come to a standstill. The municipal manager was quoted in the same article as saying 'some of these people do not even have enough water to wash, only sufficient to drink and prepare food'. (In these rural areas lack of crops and food has resulted in an increase in poaching from nearby game reserves.) With an absence of water to wash, the suffering of those with AIDS and those who try to care for them is increased.

Traditionally, bowls of water are offered to men and guests by the women of the family for washing hands before a meal is eaten. This is not possible when water sources dry up and the situation is made much worse when diarrhoea is present in the household. Although the schools at Ntandabantu now have new ventilated pit latrines, until the new rain-water tanks fill there is no water to wash hands. The results of a recent sanitation survey conducted as a part of the Masi Juleke Water and Sanitation Project (Jaffe et al. 2003: 8) in KwaNgwenya, part of the uMkhanyakude District Municipality, found that while 96 per cent of households had iron latrines after the project was completed, and while 75 per cent of these toilets were in regular use, only 45 per cent of children used them. While 92 per cent of respondents said that they washed their hands after using the toilet, water was available in only 27 per cent of the homes and only 17 per cent had soap available for use on the day the survey was conducted (Jaffe et al. 2003: 14). Cholera is an ever-present threat, made worse with the coming of rain after a prolonged drought. Dehydration after diarrhoea presents a further threat and is a common cause of death in children in the area.

Some Factors Affecting the Spread of Poverty and Disease in Rural Areas of KwaZulu-Natal

The apartheid regime undoubtedly played a role in the impoverishment of areas such as Ntandabantu. Restrictive legislation such as the Influx Control laws, which prevented African adults from remaining for more than thirty-six hours in an urban area of South Africa without a permit, both promoted and controlled the flow of adult male labour to South African mines, towns and cities. Large numbers of rural men became migrant labourers, spending most of their lives as industrial workers with short leave periods, usually about a month once a year around Christmas. Within the limited areas set aside for black Africans under the apartheid regime, a growing human and animal population exceeded the capacity of local resources to support them. In most cases, including Ntandabantu, over-use of the land triggered a cycle of degradation of rural resources.

Shortage of arable land has led to dramatic changes in subsistence production, due both to population increase and to Africans being confined to restricted areas. While some rural dwellers in KwaZulu-Natal have managed to become smallholder sugar farmers, often unsuccessful ones, the formerly tribal areas present few realistic opportunities for the production of crops or animals for sale. Land is generally insufficient for cash cropping and scarce land in tribal areas is allocated to individuals on a traditional basis, which means to men, although it is mostly women who work the land. The combination of a general demand for land and its current scarcity provides opportunities for bribery and other malpractices in chiefdoms and districts. Family lands are thus subject to progressive attrition that favours the short-term exploitation of local resources. The significance of the countryside as a source of food has declined and game animals have disappeared outside the reserves. The collecting of wild foods with high nutritional value such as the wild plum has decreased with the progressive destruction of forests and the turning of open land into pasture.

The Elderly Remember the Past

While older women noted that drought was particularly bad at the present time, they also said that rituals were performed in the past for rain. Mrs Mkhwanazi, an elderly grandmother and member of the SOFAH group, noted that women would climb a mountain in the area and pray to the goddess Nomkhubulwane for rain. 'Sometimes we even planted a plot [with crops] to thank Nomkhubulwane for giving us rain.' 'The Nazarites [or AmaNazaretha, members of a popular Zulu syncretistic church which combines Christian and traditional African ritual] used to go to the mountain. Each member had to donate fifty cents at this time. In the old days the rain used to come but now no rain comes, no matter how hard they pray.'

Another elderly participant, Mrs Mbonambi, maintained that ignoring the needs of birds and animals was also responsible for the current drought

> People nowadays don't plant sorghum. Sorghum was not only for making sorghum beer. It was also planted for the birds because they are unable to plant food for themselves. It [sorghum] was a sign to thank God for a good harvest. Sorghum was a gift to God [but was eaten by the birds]. But people were selfish; they harvested all the sorghum for their own purposes. That's why we have drought now.

A third participant and neighbour of Mrs Mbonambi, Mrs Fakude, noted

if you were harvesting, you were supposed to leave some of your harvest at the beginning and end of your plot. That left over part was for the birds. The birds belong to God but they don't plant food for themselves. So it is our responsibility to give them food as a sign to thank God for the harvest. People are so greedy [now] that they don't follow the procedures they used to in the past.

Mrs Mbonambi added that sweet potatoes, in the same way, were meant to provide food for guinea fowl. 'But nowadays people only want to plant sugar cane' a crop which has no benefit to birds. She claimed that in years gone by each household in Ntandabantu harvested not less than a hundred bags of maize, watermelons, pumpkins, cowpeas, groundnuts and other crops. Each household had its own milling machine, grinding stone or *give* (wooden pestle and mortar) to make *samp* (stamp maize). After a good harvest a family would slaughter a cow to thank Nomkhubulwane (the Zulu fertility goddess) for giving sufficient rain. In those days, she remarked, 'There were lots of tortoises but now one doesn't see a single one. Long ago, the frogs used to sing as a sign of giving thanks for rain.'

Mrs Fakude added 'nowadays we don't have *vete* [a term for frogspawn]; it was white in colour, we used to see it when we had enough rainfall'. She noted that the presence of crabs in the streams and rivers was also an indication that there was plenty of water. The women thus acknowledged the importance of biodiversity through their mention of the role of different creatures in the environment. The references to tortoises, frogs and crabs, and the reading and interpretation of natural signs in the environment connected to the presence or absence of water, were echoed by women in the village of Gcinisa, in the neighbouring province of Eastern Cape. These villagers explained that green algae growing at the edge of a local dam were called 'the blanket of the frog' (*ingumbo yesele*). 'People believe that the blanket of the frog is a house to many creatures. Frogs lay their eggs under that blanket. The blanket of the frog is a sign to us that we must look in other places for our drinking water' (Palmer 1996: 21). The presence of crabs and frogs was linked to the presence of the ancestors in the Eastern Cape and they were not removed or harmed in any way. Mrs Fakude said that when she was a girl 'The grandmothers used to tell us not to swim in dams because of the *vezimanzi* snake there.' This was a type of snake associated with the ancestors, which could be dangerous to humans. 'Nowadays there is no *vezimanzi* snake' she added. The taboo would have helped to keep the water clean.

Factors Affecting the Prevalence of HIV/AIDS

South Africa is one of the countries in southern Africa most affected by the HIV/AIDS pandemic. The prevalence of infection has been estimated to be as high as 37 per cent among 20–24-year-old pregnant women in rural KwaZulu-Natal (Harrison et al. 2000) and the prevalence of HIV in parts of the province is double the national average for South Africa (Lurie et al. 1997). The life expectancy of a male born in the Mtubatuba area is forty-two years and for a female forty-eight years. Throughout South Africa black people suffer higher infection rates than any other group and women aged 20–24 years are most affected. There are no accurate figures for the prevalence of HIV/AIDS in Ntandabantu at present. The headmaster reported that there has been a marked increase in the number of orphaned children in the primary school. When asked what happened when there was a death in the family, one participant remarked 'You get up very early in the morning and run to the *induna* to report the death. Then he signs a paper and you may bury the person.' This makes a mockery of any reliable statistics for assessing how many have died of HIV/AIDS-related causes. Community members are aware that many are dying and we are frequently distressed on our visits to hear of yet another death of an acquaintance, often a young mother. Various factors affecting prevalence are discussed below.

Many studies have shown that geographic mobility, migration and widespread population displacement are significant risk factors in the transmission of HIV/AIDS (Lurie et al. 1997). Research undertaken in Uganda showed that people who had moved within the last five years were three times more likely to be infected with HIV/AIDS than those whose residence had been stable for more than ten years (Nunn et al. 1995). Migrants were also found to have had more sexual partners than non-migrants.

Prior exposure to sexually transmitted diseases (STDs) is known to greatly increase the risks of contracting HIV. In South Africa a high rate of STD infection has been found among gold miners, and prevalence of HIV infection among migrants is estimated to be 50 per cent higher than among non-migrants (Evian 1995). Migrants' frequent and lengthy absences from home disrupt stable family relationships and the system has created a market for prostitution in mining towns. Lurie et al. (1997) estimated that 60 per cent of the households in their sample of rural Hlabisa (the district in which Ntandabantu falls) had a male migrant and 30 per cent more than one male migrant. One third of homesteads had a female migrant and another 15 per cent two or more female migrants. Female migrants in Lurie et al.'s sample lived and worked closer to their home areas than males. Although Lurie et al. noted that no women were reported to them as migrating to Johannesburg (1979: 22), one fieldworker who lives in Ntandabantu observed that many of the young people from the area, who have little hope of employment locally, borrow the bus fare to Johannesburg and soon fall into prostitution when the job they pinned their hopes on fails to materialize. He related the sad tale of his cousin, a young girl lured by the prospect of good employment in the city of gold, who returned barely eighteen months after she had left, to die 'of AIDS'.

Another participant remarked cynically that when her husband left to seek work in Johannesburg, he was likely to return with 'my present – AIDS'.

Women appear to have a more rapid progression of illness than men after transmission of the virus and present with a different constellation of opportunistic infections (Greenblatt and Hessol 2000: 8). Defilippi (2003) points out that rural women may also incur a high risk of dying from AIDS-related disease because of the contradictions in traditional cultures which promote virginity and expect married women to remain faithful to their husbands but encourage men to have multiple sexual partners. She indicates that 65 per cent of adults attending the South Coast Hospice in KwaZulu-Natal, are women, 80 per cent of whom report that they have been faithful to their partners. Approximately two-thirds of the clients treated by Anne Hutchings at the Philani clinic, a state-subsidized clinic near Empangeni, whose patients come from the same background as the women at Ntandabantu, are women in the thirty to forty age group, many of whom were abandoned by their partners when they were diagnosed HIV positive. This finding both appears to confirm the more rapid progression of illness in women and also indicates their greater willingness to come forward to be treated.

Nursing the Sick in Rural Areas of KwaZulu-Natal

In rural parts of South Africa the problem of nursing AIDS patients is becoming more acute, with a general absence of piped water, and widespread unemployment leading to near starvation in many areas. In addition, poor roads make taking the sick to hospital difficult, and there are few facilities for AIDS patients even when they do get to a clinic or hospital.

A further problem is that clinics throughout South Africa will not dispense medication to family members for a client too ill to attend personally in case it is sold. One woman noted 'the government has failed to provide the clinics with medicines. If you take a sick person to the clinic, you get Panado [paracetamol] only, and sometimes you fail to get even that, but you have hired a car to go to the clinic.' One eighty-year-old participant, all of whose children had died and who had been left as the sole support of eight grandchildren, is a known hypertensive. On our last visit we found that she had been unable to obtain her medicines because her legs were too swollen for her to attempt to walk to the nearest clinic.

In listening to older women's accounts of their difficulty in nursing sick relatives, it became apparent that an ancient view of the need for harmony between humans and their environment still held. All the informants related the pandemic to environmental ills such as drought and to the selfishness and jealousy of the human inhabitants of the area.

Cultural taboos, such as the traditional avoidance custom known as *hlonipha* or respect, between a bride and her parents-in-law, still exist and can make nursing the AIDS patient problematic, as several elderly women related. These include restrictions on

- The movement of newly wed women. A Zulu *makoti* (bride) is not allowed to enter her mother-in-law's house: 'even if her mother-in-law is sick she still has to fall on her knees at the entrance to the house'.
- Contact between men and women. Men are also not allowed to see sick women, unless they are close relatives. A man who is ill must be looked after by his wife and if he is unmarried by his parents or his brother or father. 'These relatives will take you to the men's clinic or give you traditional herbs but not to cure you of *izifo zocansi* [sexual diseases].'
- The collection or use of traditional herbs to treat patients by menstruating women.
- The use of plants incorporating the husband's tribal name. In one case, this precluded the use of *umsuzwane* (*Lippia javanica*) a particularly useful and locally abundant medicinal plant, forming a vital ingredient in the introduced creams for sores.
- The eating of certain foods. These include eggs and fish, important sources of protein for an HIV/AIDS patient who has a deprived nutritional status.

Such taboos can be overcome through sensitive interventions. Fish cakes from tinned pilchards, introduced at one of the workshops as a good and cheap form of protein, were readily eaten and one participant has since been preparing them for her family and others. She also learnt to make the *Lippia javanica* creams and supplied them to some of the sick.

The lack of medicines and essential supplies needed in home-based care of HIV/AIDS patients has been expressed by various informants. Mrs Mbonambi said that the government provides condoms but has forgotten the needs of carers for gloves, Dettol (disinfectant) and medicine such as paracetamol. She maintained that carers should be paid as their job should be done by professional nurses. The toll that the disease takes was evident when Mrs Mbonambi complained 'We are now barbarians, because since this sickness has come, one doesn't even go to church or to community meetings or attend any special occasion [ritual] in the area. If you leave your patient alone he or she will feel isolated.' Several informants said that if you nurse a sick relative, you also become ill. This is particularly the case if the patient is your son. 'If you see your children suffer serious pain, then you also feel that pain. You become weak and end up losing weight.' Mrs Mbonambi was here referring to the effects on the elderly of the burden of having to nurse severely ill relatives with no respite. However, there is some indication that elderly mothers may contract HIV from their children if they are unable to use gloves to protect themselves from possible infection from sores and infected blood. Mrs Mbonambi wanted the government 'to open hospices for our children affected by AIDS'.

Traditionally used wild vegetables known as *imifino* (a generic term for green herbs eaten on their own or used to flavour maize porridge, often translated as 'spinach' by Zulu-speakers) are an excellent source of nutrition and reach maturity more quickly than other crops. But they are reliant on available water, and, because of their small size, require more energy in collection. This can add to the problems of those whose energy store is already severely depleted by disease and adversely affect not only the women but their families. Free nutritional supplements, vitamins and limited antibiotics, antifungals and analgesics are available at some of the state hospital-run support clinics for people living with HIV/AIDS, but these are not available to the rural poor for whom such clinics are not accessible.

Water and *imifino* collecting are communal activities for women in rural areas; that is, women use these activities as an opportunity to form groups and interact with friends in a social setting for casual chat and gossip when it would not otherwise be possible for them to meet. Some of the HIV positive clients treated at the local Philani clinic have been turned away from communal water and *imifino* collecting activities because of the perceived stigma. One woman had her nose broken, while another found her only source of income lost when her status became known and parents no longer allowed their children to buy her home-made scones at the primary school. Yet others have suffered cruel rejection by their husbands' families and been blamed for the deaths of their spouses.

An added burden is that there is a great need for extra water when nursing HIV/AIDS patients. It has been calculated that a person with HIV/AIDS needs to drink at least two litres of clean water per day (Orr 2003). This is very important not only for those dehydrated by diarrhoea and night sweats but also for those suffering from urinary complications. A liquid diet is often needed for those unable to swallow because of oral thrush (candidiasis). Water is also needed for the washing of infected areas. Genital sores and open sores on the lower limbs are common and difficult to keep clean, with infection spreading rapidly to other parts of the body. In addition extra laundry needs to be undertaken to keep patients clean and comfortable. Most of the sick visited were observed to be kept very clean by their local caregivers. Considerable hardship was reported, as the following observations show:

Having to care for an AIDS patient means obtaining at least three 25-litre drums of water a day. The patient has to be bathed daily and sores dressed. If there is diarrhoea the patient's blanket and clothing also have to be washed every day.

Since my daughter became sick we have to fetch five drums (of water) per day. My son has stopped going to school because I can't do this work alone. He has to fetch water and also to cook food that the family will eat. I only cook food for my sick daughter and spend most of my time looking after her. I do her washing daily; a big towel, nightdress, underwear, and I give her a bath.

You are always tired. You have to be nice [to the patient] always. No free time for you! Some people even lose their jobs because they have to care for sick people. You must love the sick person more than your husband and spend most of your time with that person.

The mother of a family first visited in 2001, having been identified by the community as one of those most in need of assistance from SOFAH, had lost four of her five children to AIDS. Her remaining daughter later became very ill and when the authors visited the following year they found her lying on a blanket, unable to speak or move. She died six months later in February 2003. A number of teenage grandchildren also appeared to be very ill; some of them have since died, and at the last visit a six-year-old granddaughter had just been buried. The homestead was always kept spotlessly clean and swept but there was no evidence of any food resources around, except for a few chickens. A local *onompilo* (volunteer village health worker), Jabu Msimango, who was also training to be a traditional healer and is a member of the SOFAH health committee, was chosen by the committee to offer her support to the family, a task she undertook very ably.

Diet and HIV/AIDS Sufferers

Women related that special food had to be prepared for the AIDS sufferers. One carer said that you had to wake early to cook a soft porridge, which the patient could eat easily, emphasizing 'if you are sick you are like a baby, you have to drink porridge like a baby'. Such a soft porridge would be made from fresh green maize or sorghum. Traditional *isangcobe*, made from fermented dried maize or millet, provides a useful alternative, especially when fresh maize is out of season. Other foods for AIDS patients mentioned were finely chopped up watermelon, *Amaranthus imifino, maas* (curdled milk), and tea with milk 'so that he can get calcium'. Unfortunately the local water is so saline that milk often curdles when added to tea. It was also felt that the patient should be given glucose drinks 'to build up his or her strength'. These are commercially available in Mtubatuba, but expensive.

It was also noted that it was hard for rural people to get fruit (meaning citrus or deciduous fruits like apples). These are available in markets in the nearest town of Mtubatuba but are not locally grown. 'Instead we get traditional fruits from the 'forest'. These include the introduced *umdolofiya* (prickly pears, *Opuntia vulgaris*), and indigenous fruits such as *izindoni* (*Eugenia cordata*, a black edible berry much liked by children), *amahlala* (the wild Monkey apple, *Strychnos spinosa*), *ishekisane* (*Euclea udulata*) and *umdende* (a wild fig, *Ficus* sp.). All of these are only seasonally available and indigenous fruits are often small, with little flesh.

Conclusion

The provision of potable water has been a priority for the present government in South Africa. Considerable strides have been made in supplying 11 million South Africans who previously had no access to clean water. However, those who have obtained clean water have been mainly urban residents. Over 7 million, mainly rural, citizens (roughly 16 per cent of the total population) are still forced to obtain water from rivers and dams. Widespread unemployment, the advent of the AIDS pandemic, the rapid increase in TB prevalence and recent cholera outbreaks in the province as well as the severe increase in drought conditions throughout southern and eastern Africa have greatly worsened the plight of rural dwellers. In remote areas of KwaZulu-Natal, such as Ntandabantu, it is left to elderly women to care for their children dying of AIDS-related illnesses without access to adequate resources.

The absence of any regular or reliable supply of clean water makes their task almost unbearable. Children are pressed into service to collect water from dams and rivers, which often means missing school, and money spent on transport to clinics for the sick means that there is less to buy food and other necessities. Caregivers spent time and energy in making sure that their sick family members were kept clean and comfortable and bedding and clothing were washed daily. With limited resources, efforts were made to supply patients with nutritious food, cooked in a form that they could consume easily. Most of the women interviewed knew of the benefits of calcium, found in milk, and vitamin C, which could be obtained from indigenous fruits when money to buy the supplement commercially was not available. The women felt that the South African government should provide financial and other assistance to family caregivers to ease their load. Clinics were not readily accessible, were understaffed and lacked adequate supplies of medicines, often not having cheap drugs such as aspirin to relieve pain. Isolated communities such as Ntandabantu, are likely to be further marginalized in the current roll-out of antiretroviral drugs being implemented by the government. This is because the stringent counseling and adherence requirements necessitate added clinic staff and facilities. Yet, in spite of their many burdens, the women interviewed showed a resourcefulness and tenacity in their willingness to look after their sick family members, not the least of their burdens being the difficulty of access to water.

Notes

1. Names of informants have been changed.
2. The Africa Centre for Health and Population Studies is a research centre focusing on problems associated with HIV/AIDS. It is supported by the University of KwaZulu-Natal and other funders, including the UK-based Wellcome Trust.
3. A government pension, currently R782.00 per month, is paid to women over the age of sixty and men over sixty-five years, who do not have other income. This payment is often the only source of income for elderly people and families in rural areas. Payments are usually made in cash at pay-out points such as schools and clinics.

References

Defilippi, K. (2003), 'Dealing with poverty', in L. Uys and S. Cameron, eds, *Home-Based HIV/AIDS care*, Oxford: Oxford University Press.

Evian, C. (1995), 'AIDS and Social Security', *AIDS Scan*, 7(3): 8–11.

Greenback, R.M. and N.A. Hessol (2000), 'Epidemiology and Natural History of HIV Infection in Women', in J. Anderson, ed., *A Guide to the Clinical Care of Women with HIV*, Rockville: HIRSA.

Harrison, A., M. Lurie and N. Wilkinson (1997), 'Exploring Partner Communication and Patterns of Sexual Networking: Qualitative Research to Improve Management of Sexually Transmitted Diseases', *Health Transition Review*, 7 (Supplement 3): 103–7.

Harrison, A., J. Smit and L. Myer (2000), 'Prevention of HIV/AIDS in South Africa: A Review of Behaviour Change Interventions, Evidence and Options for the Future', *South African Journal of Science* 96: 285–90.

Jaffe, A., L. Dartnell and L. Torr (2003), 'Maximising Comparative Advantage in a Partnership: A Sanitation Project in Umkhanyakude District Municipality, KwaZulu-Natal', *Africanus*, 33(2): 6–17.

Krige, E.J. (1950), *The Social System of the Zulus*, Pietermaritzburg: Shuter & Shooter.

Kuper, A. (1980), 'Symbolic dimensions of the southern Bantu homestead', *Africa*, 50(1): 8–23.

Lurie, M., A. Harrison, D. Wilkinson and S. Abdool Karim (1997), 'Circular Migration and Sexual Networking in Rural KwaZulu-Natal: Implications for the Spread of HIV and Other Sexually Transmitted Diseases', *Health Transition Review*, 7 (Supplement 3): 17–27.

Mitchell, P. (2002), *The Archaeology of Southern Africa*, Cambridge: Cambridge University Press.

Nunn, A., H. Wagner, A. Kamali, A. J. Kengeya-Kayondo and D. Mulder (1995), 'Migration and HIV-1 Sero-Prevalence in a Rural Ugandan Population', *AIDS*, 9: 503–6.

Orr, N. (2003), *Positive Health, BMW*, Pretoria: South Africa (Pty) Ltd.

Palmer, R. (1996), 'Residue of Tradition or Adaptation to an Arid Environment? Cosmology, Gender and the Struggle for Water in Two Rural Communities in the Eastern Cape', Unpublished paper presented to the Association for Anthropology in Southern Africa, Pretoria.

Sansom, B. (1974), 'Traditional Economic Systems', in W.D. Hammond-Tooke, ed., *The Bantu-Speaking Peoples of Southern Africa*, London: Routledge and Kegan Paul.

11

Gender and Poverty Approach in Practice: Lessons Learned in Nepal

Umesh Pandey and *Michelle Moffatt*

Gender Discrimination in Nepal

In Nepal, a hierarchal Hindu society, stereotypical roles for men and women and a patriarchal social structure have been the main causes for the systemic domination of women. This has resulted in their very low participation in political, legislative, administrative and judicial policy-making bodies. Discriminatory laws and practices, continual dependency and lack of government social security make women further vulnerable. Though literacy and health conditions have largely improved in Nepal since the 1980s, the gender gap is still wide. Statistics on literacy vary due to differences in survey methodology, but all report a large gender gap: according to the 2001 Nepal Census women's literacy is 42 per cent compared with 65 per cent for men. Women die earlier than men in Nepal owing to high maternal mortality and the higher death rate of the girl child from 1–5 years of age. Though data reveal that women work longer hours than men and contribute more to agricultural production and household labour they are not equal partners in development with men.

However, not all women and men can be put into the same groups. Among the poorest, both men and women experience discrimination, deprivation and exclusion from public fora, while enjoying comparatively more decision-making and authority at the household and community level. Women from the Tibeto-Burman ethnic groups enjoy a more democratic social structure and play a bigger role in decision-making. They are less constrained in terms of mobility, and income-earning opportunities than women from the Indu-Aryan groups (the caste groups): 'caste, class, ethnicity, religion, geographical location and age can all be influencing factors that can change power equations for both men and women in Nepal. Therefore gender has to be carefully analysed taking all these factors under consideration to understand the existing reality' (Richardson et al. 2001: 11).

Caste, Ethnicity and Social Exclusion in Nepal

The embedded caste structure of Nepali society is directly connected with water. The so-called 'touchables' and 'untouchables' are differentiated by means of water, which has always played the role of social divider in Nepali society. The traditional view of caste is derived from the sacerdotal Hindu texts, which lay down the division of Hindu society in four orders: Brahmin (traditionally priest and scholar), Kshatriya (ruler and soldier), Vaishya (merchant) and Shudra (peasant, labourer, servant). In the Nepali context, Dalits (Kami, Damai, Sarki) are positioned in the Shudra order and have been relegated to doing work that is classified as dirty, and as a result are considered impure and 'untouchable' by so-called higher caste groups. Even today, Dalits sharing drinking water from a water point alongside higher castes is a source of conflict in many parts of rural Nepal. Higher castes have reserved for themselves the right to education, to do business and to run government.

The term 'Dalit' emerged as an identity recently in Nepal, taken from India where it gained currency in the caste Hindu riots in the early 1970s. The term, coined by the ex-untouchables themselves, is rooted in a rights perspective, claiming self-respect and dignity as an entitlement. The *Muluki Ain*, which was promulgated in 1963, legally abolished untouchability and in 1990 the constitution of Nepal also abolished discrimination and enshrined statements ensuring equality for all citizens irrespective of caste, creed or gender. However, discrimination based on caste is still a fact of life in Nepal (Asian Development Bank (ADB) 2002). The limited income earning opportunities, lack of land ownership and unequal access to education and other public resources, mean that two thirds of Dalits (12 per cent of the population in Nepal) currently live below the poverty line (Central Bureau of Statistics 2001).

Other disadvantaged groups include Janajatis, people who have their own language and traditional culture and are not included under the conventional Hindu hierarchical caste structure; they are for the most part indigenous people and constitute 35.6 per cent of the total population. They are made up of Tibeto-Burman groups who mainly reside in the hills and mountains, while Indu-Aryan groups are resident in the Terai and the hills. The government of Nepal has recognized sixty-one communities as Janajatis, and their situation varies widely from one community to another depending on factors such as physical isolation and the historical loss of traditional community-owned land. Unlike the Dalits, who face the same economic hardship wherever they live, the Janajatis are comprised of different groups of people at various stages of economic development (ADB 2002).

Gender, Poverty and Water in Nepal

Despite significant progress since the 1970s in improving access to safe drinking water, the situation in Nepal is still not satisfactory. The official national estimate for water coverage in Nepal is 72 per cent of which 71 per cent is rural and 76 per cent is urban (His Majesty's Government of Nepal (HMGN) 2002). In many parts of the country safe drinking water is still scarce and many older systems (included in the coverage estimates) are non-functioning. Fetching water from distant sources takes up considerable time and energy, and is disproportionately borne by women and girls. Even in areas where water is more accessible, the water is not always safe for human consumption. Contamination of drinking water by iron, arsenic, bacteria, and other pollutants is a problem that affects both rural and urban settlements.

Sanitation remains poor in Nepal. The official national sanitation coverage estimate stands at 25 per cent in Nepal of which 21 per cent is rural and 53 per cent is urban (HMGN: 2002). The impact of poor sanitation on the health and well-being of women, men and children is significant. The increased work burden for women, who are the main caregivers, and the decreased productivity caused by illness are particularly costly for the poorest, since they cannot afford to lose their daily wages.

Women are the main collectors and managers of water at the household level, and have a major stake in maintaining water supply systems. With improved systems women save time collecting water; for example, in NEWAH projects the average time saved is two hours per day due to water points being nearer homes. Women can use this time for productive purposes, other household chores or much needed rest. Acknowledging these benefits has led government policy-makers, donors, and aid agencies to a commitment to increasing the involvement and participation of women in water and sanitation projects. However, in reality, while women are usually involved in project implementation, they often lack access to decision-making in the planning, designing and management phases. The poor are even further marginalized due to their low power, status and capacity to express their concerns. The participatory approaches used to involve the 'community' assume that the community is homogeneous. They do not take into account the different degrees of power and influence that exist, the needs and concerns of rich and poor women and men, and the economic losses accrued for the poorest by participating in project activities as unpaid labour (NEWAH 2004).

Definitions and Approaches Regarding 'Gender Mainstreaming'

The international debate has moved from women in development (WID) to gender and development (GAD) or gender mainstreaming, recognizing that keeping the activities and projects for women in isolation does not address the issue of gender roles in society. Gender mainstreaming is seen as a means of addressing gender discrimination, and commonly cited criteria for gender mainstreaming are:

- The integration of gender equality concerns into the analysis and formulation of all policies.
- Initiatives to enable women as well as men to formulate and express their views and participate in decision-making across all development issues.
- Initiatives specifically directed towards women. Similarly, initiatives targeted directly to men are necessary and complementary as long as they promote gender equality (OECD DAC 1999: 12–13).

'Gender mainstreaming and social exclusion are linked in that they both address the need for inclusion in society whether it is between men and women or between different groups' (Richardson et al. 2001: 14). Current literature on gender mainstreaming often highlights management support as a vital ingredient. In NEWAH's view, this is essential to kick-start the process of mainstreaming; without support from managers, institutions consistently fail to make any progress.

Gender and Poverty Mainstreaming in NEWAH

Mainstreaming a Gender and Poverty (GAP) approach in Nepal Water for Health (NEWAH) is an ongoing process, which includes working at programme and organizational levels, sharing lessons learned and identifying the challenges ahead. NEWAH's key partner at the start was WaterAid Nepal (WAN) and the first country representative was very supportive of this gender mainstreaming initiative and hired a gender consultant to help to develop and implement the GAP approach. Funding for this was subsequently obtained from DFID in 1999. The next WAN country representative did not prioritize gender and did not support WaterAid Nepal after the first year. NEWAH considered this to be a missed opportunity for WaterAid to implement the gender principles that it advocates at global level. The process was therefore entirely led by NEWAH, supported by the consultant and funds from DFID.

NEWAH began its mainstreaming process in late 1998 with discussions involving senior management staff about gender and poverty. A common understanding was built, defining gender inequality as a major constraint to poverty reduction. To address it NEWAH considered tackling social exclusion and promoting gender equity as cross-cutting themes in all their work. Strong links between gender, caste, ethnicity and poverty were identified, and the analysis of poverty included identifying groups that were vulnerable to poverty and social exclusion; gender disparities were understood as a particular issue. A number of specific problems were highlighted. In the community Water User and Sanitation Committees (WUSCs) NEWAH's experience was that the richest, higher caste men dominated all aspects of the project and women and poor men were often not represented in any decision-making or training related to improved water and sanitation systems. NEWAH had previously experimented with all-women WUSCs, but these were not entirely successful, since men often did not support the all-women committees that were imposed on them.

Domination of water systems by male elites in Nepal often leads to unequal access to safe drinking water between the better off and poorer socio-economic groups and ultimately to the unsustainability of projects. NEWAH's experience in two rural communities illustrates this. At least two gravity flow water systems in the Himalayan hills were suffering from decay due to illegal pipe connections made by the higher caste elites at the head of a gravity flow system. This led to unequal access to improved drinking water by Dalits and other socially excluded ethnic groups residing at the tail of the system. The situation resulted in these households being unwilling to pay monthly user fees for operation and maintenance.

Further analysis was continued by NEWAH in a 'Gender Awareness for Poverty Reduction' workshop for senior and middle management staff in early 1999, which proved to be a catalyst for substantial change within NEWAH and its programme. An essential ingredient for the success of this crucial workshop was the ability of the gender trainer from Save the Children US, who delivered training in Nepali and was highly skilled in facilitating a practical and participatory learning agenda for participants. The model used was SCF's Gender Relations Analysis framework, which included a practical guide to mainstreaming at institution and programme level. This was critical because there is a tendency in institutions to highlight gender as an issue in the analysis of poverty, but to ignore the importance of institutions and the formulation of strategies in addressing gender inequality issues. The workshop developed a clear action plan and process for the future. Another asset to the workshop was a powerful presentation of PhD research findings (Regmi 2000) on gender and institutions in the rural drinking water sector in Nepal. Management staff agreed that having a man strongly advocate gender equity certainly added value to the learning in this initial workshop.

The Process of Institutional Change

Following the workshop, NEWAH's management began to review personnel policies, working principles and their strategic plan from a gender perspective. Senior management took responsibility for this task over the next two years. The outcomes of a personnel policy review included increasing paid maternity leave, provision of childcare allowance, introduction of one week's paternity leave and funeral leave, and targets for recruiting more women into the institution. The benefits of the review accrued to both men and women; this was important in reinforcing the message that a focus on 'gender' was not about improving working terms and conditions for women only – men were targeted also as part of a family-friendly personnel policy.

As a result of personnel policy changes in 2000, the number of women staff in NEWAH increased from 13 to 24 per cent in four years. There are now fifty-seven women out of a total of 240 staff; 17 per cent of senior-level staff members are now women with other women working at the mid and lower levels. In its strategic plan NEWAH committed itself to increasing the number of women to 25 per cent by 2005 and may well exceed this target. However, there are constraints to increasing the number of women employed in the rural drinking water sector in Nepal, since there is only a very small pool of technically trained women. Women are not encouraged by parents to enter into technical fields and marriage often restricts the mobility of women, which is frequently required in the sector. Most women are therefore employed in the more socially acceptable health and sanitation components of projects. NEWAH has increased the number of technical on-the-job training posts to enable more women to train for technical positions and in 2004 women made up 38 per cent of technical on-the-job trainees in NEWAH.

Senior managers led the implementation of the Gender and Poverty approach. Despite gender-awareness training it was disappointing (but not entirely un-expected) to observe in the early stages that some managers, both men and – more surprisingly – women, were insensitive to women's issues; this was challenging for the organisation. NEWAH wants to increase the number of women staff, especially at senior levels, and is considering how a human resource development strategy can enable women as well as men to first develop their gender awareness, skills and competency, before taking up senior decision-making responsibilities in the future.

The GAP process has enabled NEWAH to discuss openly how to ensure greater diversity in its caste/ethnic staffing ratio to reflect diversity in the larger society. Social exclusion is a strongly sensitive issue in Nepal and there is resistance to change in this area; challenging it needs to be carefully implemented over a long time frame. There are still areas for improvement; NEWAH's Board comprises

29 per cent women and this needs to increase. Similarly, the Board has begun to discuss how to attract people who represent caste and ethnic diversity and the poor. NEWAH recognizes it is important that its governance not only supports but also demonstrates what it believes, but this has been a challenge in practice and takes time to achieve. However, among all castes and ethnic groups recruited as new staff between 1999 and 2004, the biggest increase has been in the Dalit group (from 3 to 7 per cent). While there is still room for improvement this indicates positive progress.

NEWAH approaches diversity as an issue of social exclusion. There is now recognition, for example, that lack of field staff who can represent and speak the ethnic language of a community presents a barrier to communicating with men and, particularly, women whose first language is not Nepali. NEWAH often has to rely on district partner NGOs to undertake translation, which presents additional challenges for project implementation. In many of the communities surveyed by NEWAH (particularly in the Terai and the far west), Nepali is not the mother tongue and understanding and speaking Nepali can be especially difficult for women. The high illiteracy rate among women and the poor also needs to be taken into account; staff are needed who can speak the local language and use tailor-made materials, including pictures for those who cannot read. This will help to ensure understanding of training and project-related activities, particularly health and hygiene education (NEWAH 2004). NEWAH has recently begun to tackle this issue by hiring staff from the Kamaiya, very poor ex-bonded labourer communities, some of whom NEWAH serves.

The Process of Programme Change

At the programme level, a Gender and Poverty (GAP) Unit was established in 1999 comprising thirty-two technical and social staff. Five regional GAP teams and one team based at headquarters were created. A period of training followed on gender awareness and training of trainers, as well as applying Participatory Rural Appraisal (PRA) methods gender-sensitively. GAP teams then delivered gender awareness training to peers in NEWAH, NGO partners and communities in five pilot projects. This was the beginning of the establishment of a critical mass of operational and middle management staff within NEWAH as gender trainers and advocates to take gender mainstreaming forward. During this time a strategy was developed and the teams began, in 2000, to pilot what became known as a 'GAP approach' in five villages – one in each of the five regions of Nepal – with a total population of 5,544.

In formulating this approach the GAP teams felt that the questions of who could participate and who was included were crucial in addressing poverty reduction.

One team began to address these issues head-on at the feasibility stage of a GAP pilot project. The project had been requested by a local politician, who already had a tap stand close to his house and who had not included a poor and needy Dalit community in the plans for the project. The team liaised with the government district line agency, the Dalit community and the politician to ensure that the community were empowered to access and manage the project. This took time and numerous meetings, which often ran late into the evening. It tested the confidence of the team to negotiate on behalf of a powerless community to obtain changes, but the team felt it had begun to address a significant inequity, which is all too familiar in Nepal: that of inequitable access and control of water resources between the rich and poor.

GAP teams reported that the pilot projects presented them with major challenges, especially in facilitating discussions on sensitive gender and poverty issues. A very real concern for staff at the beginning was addressing initial resistance by elites and how to avoid conflicts developing. A fear of failure was felt among a number of staff. This fear should not be underestimated at the community level, where in many villages gender and caste discrimination is deeply embedded. Recently, for example, staff conducted an assessment in a community on behalf of another agency (NEWAH 2003). After one meeting a man was observed brandishing a bunch of nettles doused in cow urine, with which he proceeded to sprinkle the men to 'purify' them, since a menstruating woman had apparently attended the meeting. This example is one of many experienced by development workers in Nepal.

One of the overriding constraints is the negative attitudes that are deeply rooted in Nepali society concerning the capacity of women. In some GAP project communities, men were initially reluctant to rely upon women for tasks that were traditionally not done by them, such as caretaking water schemes in hill projects. This attitude was also initially apparent in NEWAH when women staff began to undertake more senior and challenging responsibilities. However, the fact that no conflicts were experienced within NEWAH or the pilot communities and that small but very significant initial gains have been achieved has considerably increased the confidence of GAP staff over the past five years. Both men and women in the teams feel they now have a common understanding, work together better as a team and believe they can sow seeds for positive social change.

The NEWAH GAP proposal was submitted to and accepted by DFID, Nepal in 2000. DFID is increasingly committed to ensuring that women as well as men benefit from development work, as outlined in *Poverty Elimination and the Empowerment of Women*: 'the purpose of DFID's strategy is to ensure that women's empowerment and gender equality are actively pursued in the mainstream of all development activities' (DFID 2000a: 28). DFID has addressed social exclusion in its paper *Realising Human Rights for Poor People*. As an operational principle underpinning a rights perspective in development, social inclusion is defined as 'building socially inclusive societies based on values of equality and non-discrimination, through development which promotes all human rights for all people' (DFID 2000b: 7). NEWAH's GAP approach was seen as a potential vehicle for achieving DFID's goals around these strategies in the rural drinking water sector. In the summer of 2002, as part of a phasing-in of the GAP approach throughout the programme, NEWAH trained more teams as gender awareness trainers. This enabled a further building of in-house capacity to implement, monitor and evaluate gender and poverty projects. Without this capacity NEWAH would not be able to deliver gender awareness training to the fifty to sixty new communities and local partners it works with each year. In 2002 the GAP approach was integrated in around 35 per cent of NEWAH's programme, while 100 per cent integration was achieved in 2004.

Features of the GAP Approach and Community Impact

Well-Being Ranking to Identify the Poor
To be poverty-sensitive it is necessary to identify the poorest. A Well-Being Ranking PRA exercise was introduced in the pilot projects in order to target the poorest households for subsidies. Local households themselves decided the criteria for socio-economic ranking; criteria included amount of land ownership, food sufficiency, employment/income, indebtedness such as bonded labour (now outlawed) and disability. In the pilot projects most often the Dalit households filled the lower socio-economic groups, but not exclusively. However, there were no Dalit households in the better off groups, reflecting the link between caste and poverty. The results of the exercise were presented in a mass community meeting and debated until a consensus was reached. Identifying the poorest households enables NEWAH and communities to provide them with additional support such as paid labour, subsidized latrines, and differentiated financial contributions for operation and maintenance (O&M), in agreement with the WUSC.

Graded Rate System of Operation and Maintenance (O&M) Payments
The GAP approach recognizes that flat rate user fees for financing water supplies penalise the poor, who have to spend proportionally more of their income than the better off. The poorest, including widows and divorced women without cash income from other family members, are often the worst affected. A graded rate maintenance system for O&M was piloted with two communities that allowed for the poorer groups to pay less than the better off. Despite initial resistance by some better off households, they were subsequently persuaded to participate in the system on the basis that the poor could only pay what they could afford; reducing

defaulting would increase sustainability of the water system. Two years after the system was initiated, one community had revised its rates, but still retained the graded rate system. Tea shop owners who previously had to pay for water to be transported, now have a regular supply and are willing to pay proportionally more than domestic users, because it is still cheaper than transporting water. This system has led to more regular payments by users, with stronger action against defaulters. The GAP team, propelled by this initial success, has initiated the system in another ten communities in the far west of Nepal. NEWAH will continue to monitor this new systems of payment in the long term, for its impact on willingness to pay and project sustainability.

Payment to the Poorest Men and Women for Unskilled Labour Contributions

The GAP approach recognizes that poor men and women are often 'volunteered' by elites to contribute unpaid and unskilled labour in water and sanitation projects. NEWAH projects sometimes require up to fifty free labour days from each household as part of their community contribution, depending on the number of households and their distance from roadheads, during construction of the water system. This means that poor families are unable to work in their fields or as daily wage farm labourers during this period, resulting in a loss of vital income. The projects therefore introduced payment of 50 per cent of the standard daily labour rate (around 60p per day) for the poorest household members (both men and women) who contributed unskilled labour. The evaluation confirmed that all socio-economic groups viewed this as a positive intervention. There were initial concerns that this cash income would be misappropriated by men and spent on alcohol and cigarettes; one community WUSC was so concerned that it decided to pay in the form of rice. However, these concerns proved to be unfounded and cash was reported to have been spent on food, domestic needs and in some cases income-generating commodities such as chickens and goats. The provision of these payments for the poorest households led to an additional minimal cost to the project of 1.4%.

Priority for Technical Training and Paid Technical Jobs for Women as well as Men

The GAP approach places emphasis on paid job opportunities for women as well as men. Traditionally male elites have come forward for training and paid job opportunities, although it is women users who often notice any water system defects first as they visit the system every day. In the pilot projects women as well as men were trained as system maintenance caretakers, enabling women to swiftly respond to the breakdowns that directly affect them, especially when men are away as seasonal migrants or as a result of the ongoing Maoist conflict. It is frequently the case in rural water projects that a trained male caretaker leaves with his newly acquired skills to undertake a better-paid job in a town. Women trained as caretakers were paid during the project but were not employed subsequently. Following the project, although they performed as competently as their male colleagues, and were sometimes allowed to help out, this was only on a voluntary basis. However, all the trained women reported a rise in their self-confidence and social status both at home and in the community, and they could use their skills to respond immediately to breakdowns when necessary. The issue of access to paid technical jobs for trained women post-project will be considered by NEWAH in the future, by linking these women with other development programmes in these communities.

Women were also trained as sanitation and water system masons, along with men. Some women were initially resistant, because they feared being ridiculed by the community. Confidence and awareness building enabled most women to train and become effective in their new roles, with the support of the community. However, in one majority Muslim community a woman was trained as a sanitation mason, but the contractor would not accept her in her role and she discontinued. This was disappointing for GAP staff, but it was a lesson learned. It brought home the complexities of working to achieve social change in communities with deeply embedded gender discriminatory practices.

GAP teams learned that particularly intensive preparation time is needed in some conservative communities to allow women to take roles that are seen as strictly the domain of men. A team working with a Muslim community was able to encourage women to attend meetings, which previously was unheard of. In this community it was difficult for women to openly participate in meetings due to the system of *purdha*, where women should not show their faces to men. Several house-to-house visits and meetings had to be conducted, with discussions in the local dialect, to negotiate with men to allow women to attend meetings. As a result WUSC members reported that father-in-laws began to converse with daughter-in-laws, which was previously taboo. Even modest gains and breakthroughs of this sort were significant.

Gender Balanced Community Water User and Sanitation Committees (WUSC)

The pilot projects aimed to achieve a 50/50 gender balance in community management committees, to help ensure women's participation in decision-making over water resources, and to minimize the domination of management committees by male elites. Many agencies in Nepal recognize the need to empower women to participate in decision-making and many follow sector policy that recommends at least two women be elected to committees. Often this proves to be disempowering, with women being elected by men without their knowledge, or without the training or confidence building needed to make them effective in their roles. Few agencies place emphasis on the importance of facilitating the formation of WUSCs, yet this is essential to build a critical mass of women to take decision-making roles on executive committees.

GAP teams were pro-active in persuading communities of the benefits of active mixed gender committees for democratic decision-making. While this was a challenging task, the results of the five pilot projects are encouraging and the pilot villages had more women in the WUSCs and in key management positions than other villages. Women were re-elected into the same positions after two years. An equally important achievement is the increased representation of poor Dalit and ethnic men and women in some of these committees, although only a few were in fact elected from the poorest households. In GAP villages the poor and Dalits could vote; in non-GAP villages only better off men, and in some instances women, decided representation. The lack of participation in decision-making by these marginalized groups is reflective of many communities in Nepal. Future project design will need to consider how to further increase the participation of the poorest and marginalized groups, especially in decision-making.

NEWAH has been discussing the issue of sustaining management skills within the WUSC, since re-elections occur in WUSCs every two years as set out in the WUSC constitution, which is standard throughout the sector. Yet newly elected WUSC members who replace those originally trained do not have access to management training, after a project has been successfully implemented. There is an obvious need to provide longer-term support to WUSCs, but currently no agency in the sector, including NEWAH, provides this. The question is how NGO agencies should provide longer term capacity-building support and for how long? NEWAH believes that this type of support must be provided by NGOs, at least until government structures are in place to offer it.

Transforming Gender Roles within the Household

The GAP teams delivered training in gender awareness and the new approach to both partners and pilot communities. The purpose was to raise awareness of how gender-prescribed roles and attitudes have negative impacts on the family, community and development. A change in gender roles was analysed during the GAP evaluation and it was found that men in GAP villages helped out with common household chores more than men in non-GAP villages, allowing women to save some time and effort and to increase their participation in community decision-making activities. Women in GAP projects spent less time collecting water than the women in non-GAP projects, largely because other members in the household contributed more time and effort to this shared household task (NEWAH 2004).

Although the share of household tasks has increased for men in the GAP villages, women still carry out the majority of tasks. Boys on the other hand continue to have the smallest share of household tasks among family members reflecting the societal practice of discrimination against girls. The socialization process of gender roles starts from a young age, so promoting gender equity between girls and boys will be further stressed in future gender awareness training and school programmes at the community level. Increased equity in household decision-making was found in GAP project households; women in GAP villages reported increased control over the use of their own income, whereas more women in non-GAP villages reported being powerless to stop their menfolk from spending their money (NEWAH 2004).

Health, Hygiene and Sanitation

As part of the GAP approach men as well as women have access to health, hygiene and sanitation education. Men need to be encouraged to change personal hygiene behaviour and, in their role as fathers, to assist their children in hygiene practices such as hand washing after defecation and before eating. If every member of the family has access to increased health, hygiene and sanitation knowledge, the chances of changing hygiene and sanitation practice are higher and the positive impacts on health will be greater. The need for such access is reflected in the appalling statistic of around 15,000 children under the age of five dying from diarrhoeal-related disease in Nepal annually (His Majesty's Government of Nepal 2004). The national sanitation coverage estimate in 2002 was 21 per cent in rural Nepal, and poor hygiene and sanitation remain a fundamental cause of poor quality of life in rural Nepal. In projects men are often resistant to joining women in health education, as they associate health, hygiene and sanitation with the traditional role of women as caregivers. NEWAH recruited gender-trained male and female community health motivators in each project; the male health motivators were able to persuade more men to attend health, hygiene and sanitation classes and to discuss health and related gender issues.

However, when the impacts of changed hygiene practice were evaluated in 2003 no major differences between GAP and non-GAP villages in health and hygiene behaviour were found, though there was a marked difference between GAP and non-GAP villages in the Terai region. It was reported in one village that while men were interested in attending health and sanitation classes with a male health motivator, the classes were held during the time of water system construction, when men were busy labouring on the project. Hence men attended some, but not all, classes. The health motivator noted improved hygiene behaviour during project construction but after the project the changes declined. This has led

NEWAH to consider increasing the number of health and hygiene classes over a longer time period, to target more men, as well as women, and to increase the impact of the training.

There is still a lack of adequate funding for health, hygiene and sanitation promotion in Nepal. If changes in hygiene and sanitation behaviour are to be sustained, then greater investment is needed. The relatively minor additional costs to projects in hiring male as well as female motivators and increasing the number of training sessions for a longer period to ensure better access are justified if poor sanitation, health and hygiene practices can be improved. It is estimated that Nepal is losing approximately 9 billion rupees each year as a result of poor health (His Majesty's Government of Nepal 2002), related to poor hygiene and sanitation.

There was a notable difference between GAP and non-GAP villages in relation to sanitation coverage. The percentage of poor households with toilets in GAP villages was nearly twice as high as in the non-GAP villages. These results are significant and can be attributed to the GAP approach of targeting women, men and children in its hygiene promotion programme and providing free latrine (toilet) components for the poorest households. Subsidies are strongly argued against by some in the sector, but they are entirely justified for the poorest who live below the poverty line. Experience shows that even when the poorest are convinced of the need for a toilet to improve hygiene and sanitation, they simply do not have the cash to invest in a toilet. To date various approaches have not yet been able to adequately address the issue of the inability to construct a latrine due to lack of cash or landlessness among the poorest. NEWAH is now piloting a new community-led Total Sanitation Approach, which will focus on more intensive awareness-raising rather than subsidy; it will include the formation of separate community action committees represented by men and women from all castes and ethnic groups in a community.

NEWAH has been implementing a Child-to-Child health and sanitation approach in schools for six years. In the GAP pilot projects this approach was expanded to 'out-of-school' children. The Child Health Awareness Committees were comprised of an equal number of boys and girls who trained both 'in-school' and 'out-of-school' children in health and sanitation education via school curriculum, posters, competitions and street theatre. The impacts of this approach have yet to be evaluated, but anecdotal evidence suggests it was very successful, as children were able to act as agents for change. For example, some were able to apply pressure on parents to purchase and use a latrine. Perceptions of teachers and parents revealed there was improved personal hygiene in 75 per cent of schools and improved school environment in 85 per cent of schools. NEWAH is considering expanding its child-to-child health and sanitation approach as it believes that perhaps children's attitudes relating to poor hygiene and sanitation are not so entrenched and can be more easily changed than those of adults. In particular NEWAH believes that targeting teenagers in secondary schools may have a significant impact, since in a few years they will be married with their own families to educate and care for in relation to hygiene and sanitation.

Equitable Access to Water

The Caste Debate
Like other implementing agencies in Nepal, NEWAH has a flexible policy relating to the number of tap stands per cluster of households. In situations of caste conflict that cannot be overcome, tap stands are built separately, particularly for geographically isolated and deprived Dalit households. While this issue is politically sensitive and some argue provision of separate water points reinforces caste discrimination, such position does ensure equal access to safe drinking water. Caste discrimination cannot always be addressed in all communities in the short term, but equal access to improved water can be addressed, regardless of caste or class. In one GAP project a small remote cluster of three Dalit households opted for a smaller tapstand to reduce their contribution in local materials and because they did not require a full-size tapstand. In another project, clusters of high-caste and Dalit households agreed they should and could share a water point.

The approach taken by NEWAH demonstrates that 'one size does not fit all'; it is not appropriate to assume that all projects can and should address gender discrimination and social exclusion equally. This issue has been strongly debated between NEWAH and one of it's donors, who questioned whether it would be appropriate for NEWAH to cancel a project in a community if agreement to share tap stands between higher castes and Dalits could not be agreed. NEWAH's stance is that in some remote and conservative communities caste discrimination will take a long time to be eradicated, despite everyone's best efforts. In the meantime, NEWAH believes its responsibility is to ensure everyone has the right to equal access to safe drinking water, especially the poorest who live below the poverty line.

Technology Design Modifications
NEWAH's experience is that consulting women as well as men on technology design options ensures technology is appropriate to meet both men and women's practical needs. Under the GAP approach women are consulted over the number, location and orientation of tap stands and tube wells to ensure their needs are met. For example tap stands that have to be located by a busy trail have, at the request of women, been positioned facing away from the trail to increase privacy. Women were consulted on the height of parapets used by them to wash clothes and related

drainage requirements. Through consultation, some GAP communities have opted for community-financed communal bathing units, which have increased bathing. As one woman user commented, 'bathing is now a pleasure for us as we don't have to worry about men seeing us'. Tap stands in schools are now constructed with lower taps to allow smaller children easier access to drinking water.

Latrines built in schools are designed to be more child-friendly to address the needs of girls. Teenage girls need more time for latrine usage than boys, therefore the numbers of units for girls have been increased and they have been designed to meet girls' sanitary needs. This is expected to increase the attendance of secondary-level girls, in particular during menstruation, when girls stay away from school because of the lack of latrine facilities. In one GAP project, a secondary school teacher mentioned that he noticed the improved attendance of girls due to the provision of latrines specifically for them. NEWAH is also reviewing latrine designs for adults to meet the specific needs of men and women, taking into account, for instance, difficulties with sitting in a squatting position due to pregnancy, old age and disability.

Lessons Learnt

Through the GAP evaluation a number of lessons were learnt, valuable for both NEWAH and its communities, as well as the sector.

Long-term Planning Needed to Promote Equity and Poverty Issues

Addressing gender and caste/ethnicity issues is a sensitive topic in a society that continues to uphold and practice discrimination based on gender, caste and ethnicity. It requires not only changing entrenched social, religious and cultural beliefs and values but addressing unequal power relations. The GAP approach recognizes that long-term intervention is needed to change behaviour at the organizational, community, household and individual levels, and requires an understanding of the causes behind social inequity. Changing behaviour is achieved by building the capacity of staff to integrate social equity issues at all stages of the project cycle. This includes capacity-building of project partners and communities through gender and poverty awareness training, with the understanding that changes cannot happen overnight.

Organizational Commitment

Having a gender strategy is an important first step; however, commitment at the organizational level in terms of allocating adequate financial and institutional support is also important. Much of what has been accomplished within NEWAH and the project communities would not have been possible without the support of NEWAH's Board and its management. It is recognized also that the impetus of change should not remain solely with management, so NEWAH makes it the responsibility of all staff through the formation of operational teams of trained GAP staff members.

Gender- and Poverty-sensitive Indicators
Mechanisms need to be in place to measure what progress has been achieved at the organizational level. While NEWAH's evaluation methodology does this at the project level, NEWAH will consider developing systematic indicators to assess changes over time in staff gender awareness, attitude and behaviour towards gender, diversity and poverty issues. Anecdotal evidence and experience indicate a positive change in awareness, attitude and behaviour has occurred to a degree unexpected at the outset.

Empowerment of Women and Socially Excluded Groups
The GAP approach has increased women's ability to have voice and choice in the location, design and orientation of water points. Some poor women have also benefited from being trained to undertake paid technical jobs during project implementation, and have achieved increased income and social status within their households and communities. Executive positions on WUSCs have enabled women to influence decisions on the management of water and have increased their social status. GAP villages had a relatively strong representation on the WUSCs from different socio-economic groups; in the GAP projects the majority of female members of the WUSCs belonged to the disadvantaged castes, ethnic groups and religious minorities, whereas in the non-GAP projects the majority belonged to the higher castes. Women in GAP projects increased their participation in meetings and community activities. In strong contrast, the women in the non-GAP projects reported that while they attended meetings they lacked influence in decision-making or did not actively participate.

Increased equity in household decision-making was reflected in the increasing control that women reported over the use of their own income. Both the gender awareness training and the focus of the GAP teams on getting women positions in the project appear to have positively influenced both men and women in these villages, giving women greater confidence and ability. While the GAP approach has certainly had an empowering effect on a number of women, as well as men from socially excluded groups, it cannot guarantee that these effects will be sustained and replicated in the longer term in these villages. The GAP approach is 'work in progress', which will continue to look at ways of increasing and sustaining positive change in gender relations in communities, and monitoring progress.

Sustainability

This assessment only examined water supply schemes up to two years after project completion. Although the water points in the GAP villages were better maintained and more reliable than those in non-GAP villages, it is not possible to predict what their condition will be like in the future. NEWAH may consider extending the GAP pilot post-project evaluations up to five years, to better assess the sustainability of the GAP approach.

Conclusion

NEWAH as a changing institution is also shaping its partner organizations in promoting gender-sensitive pro-poor policies and implementing them at community level. In the GAP projects, NEWAH has oriented partners about the rationale of the approach and provided GAP training and skills at community level for successful implementation. It is too early to say whether the approach will be successful in contributing to long-term sustainability of these projects, but up to two years on, the indications are promising. NEWAH is now in a position to influence the behaviour of up to sixty partner organizations and communities with whom it works annually. It is also informing its long-term donor partners that a serious commitment to gender mainstreaming and social inclusion, with appropriate resources, can bring positive results.

Scaling up gender- and poverty-sensitive community management of water supply and sanitation systems requires institutional commitment and resources for long-term support. It is essential to view women and men in a community as equal, regardless of caste, ethnicity, class or religious factors, and to respond to their needs. This important dimension is too often neglected in all stages of a project cycle in the water sector in Nepal. NEWAH has begun to play a key advocacy role promoting incorporation of gender and poverty issues (NEWAH 2003), by demonstrating the need and sharing its practical experiences. All the indications are that addressing gender and poverty issues can substantially contribute to achieving equitable access and sustainability for poverty reduction. What has been significant in NEWAH's case is its ability to change institutional attitudes towards accepting and implementing gender-sensitive and pro-poor community management. This takes time, but NEWAH's openness to learning and reflection is illustrated in its *mantra* 'start from within, practice what we preach', and it has shown in practice that gender mainstreaming can be translated into practice and have far-reaching effects.

References

Asian Development Bank (2002), *Poverty Reduction in Nepal – Issues, Findings and Approaches*, Kathmandu, Nepal.

Central Bureau of Statistics (2001), National Census, 2001, Kathmandu, Nepal.

DFID, (2000a), *Poverty Elimination and the Empowerment of Women*, Strategy Paper, London: DFID, UK.

—— (2000b), *Realising Human Rights for Poor People*, Strategy Paper, London: DFID, UK.

His Majesty's Government of Nepal (2002a), HMGN 10th Five-Year Plan, Kathmandu, Nepal: National Planning Commission.

—— (2002b), National Sanitation Policy, Kathmandu, Nepal: Department of Water Supply and Sewerage.

—— (2004), National Hygiene and Sanitation Policy, Kathmandu, Nepal: Department of Water Supply and Sewerage.

NEWAH, (2003), Consolidated Report of a *Socio-Economic Survey for the ADB Project Preparation Technical Assistance of the Community-Based Water Supply and Sanitation Project*, Kathmandu, Nepal: Nepal Water for Health. Unpublished.

—— (2004), A Summary of Evaluation Findings of NEWAH's GAP Approach Using the NEWAH Participatory Assessment (NPA), Kathmandu, Nepal: Nepal Water for Health.

OECD DAC (1999), *Guidelines for Gender Equality and Women's Empowerment in Development Co-operation*, Paris: OECD.

Regmi, S. (2000), 'The Management of Gender Issues in Water Projects in Nepal', Unpublished PhD thesis, University of Southampton, UK.

Richardson, L., R. Thapa. and B. K. Shrestha (2001), *Social Exclusion and Gender Study*, for DFID Nepal. Unpublished.

UNICEF/HMGN (1999), *Basic Sanitation Package*, Kathmandu, Nepal: Department of Water Supply and Sewerage.

12

Easier to Say, Harder to Do: Gender, Equity and Water[1]

Sarah House

Introduction

The importance of considering gender in water projects has been openly talked about at international fora since the 1970s and national domestic water supply and water resource policies have increasingly incorporated considerations of gender issues, although to differing degrees and with varying approaches (Gender and Water Alliance (GWA) 2003: 75–83). Although it would have been hoped that a greater impact would have been made on policies worldwide after this time, it is disappointing to notice that the policy and practice at organizational and project levels in international NGOs and organizations working in the water sector remain erratic and in some cases still manage to ignore the considerations altogether.

This chapter looks at some of the challenges that have led to this gap between international policy and organizational policy and practice. The experiences are written from the perspective of the work of an international NGO (INGO) and partners in northern Tanzania.[2] The chapter relates to the practical work undertaken on gender and equity by the KINNAPA/KDC/WaterAid Kiteto Water, Hygiene and Sanitation Programme in northern Tanzania, a collaborative programme between WaterAid Kiteto, the Kiteto District Council, and a local NGO, KINNAPA Development Programme. It covers the period 1999–2002, and relates also to experiences within WaterAid Tanzania and WaterAid as an international organization. Within the programme equity was in particular considered between people from different ethnic groups who had differing livelihood bases of agriculture, pastoralism and hunter-gathering.

WaterAid is a thinking and learning INGO which is open to new ideas and responsive to learning from its partners and programmes. It has numerous successful projects and has had wide impact within its programme areas around the world. The difficulties faced in considering gender and equity within such an organization can only put into perspective the challenges which may be faced in getting less reflective organizations, or more established agencies or institutions, to mainstream gender in a practical way in their policies and programmes.

Experience has shown that the importance of considering gender and equity in water, hygiene and sanitation projects has not always been understood, or agreed upon within WaterAid. But experience has also shown that when the opportunity has been available to make progress on gender and equity there have still been many challenges to face.

Getting Debates Started

Getting Started and Views on the Consideration of Gender

The consideration of gender in projects still causes significant debate and differences of opinion. There seem to be differing feelings and understanding about what considering gender means and whether or not we should be considering it. Typical comments range from the assertion that gender is a Western or feminist agenda not suitable for use in the South, to 'if we get some women on the water committees and digging trenches then we've succeeded', to 'it means working to change the structure and power relations of society so that men and women are on equal footings'. This in turn means that you can not do water projects on their own, that they must be supported by other activities such as literacy and income generating activities which will help in the strategic empowerment of women.

In Kiteto, for reasons explained below, little effort was made initially to overcome the many barriers to women's participation.

> On arriving in the first village on taking up post in Kiteto, we sat down with a number of village members and our (usually) all-male team and had a meeting essentially between the male village chairman and the programme and partner staff. The programme had been working in this village for over two years. On questioning why there were no women at the meeting and how we could go ahead with a meeting without them, a management-level staff member replied 'well when women do come they don't speak, so what is the point?' (House 1999, Kiteto, Tanzania)

In Kiteto, WaterAid was behind local partners in interest, understanding and desire to consider gender in the work. Both KINNAPA, the local pastoralist NGO, and the Kiteto District Council (KDC) were making various efforts to look at themselves and the aspects of their work which were funded by donors other than WaterAid, although at this time they may not have known how to respond practically to the issues. Many of the gender issues in the district were stark and the local NGO faced a number of these on a daily basis. The fact that the local partners, interest in gender issues had not spilled over into the WaterAid-funded programme seemed to stem partly from the power relationship between WaterAid and its partners, capacity issues in general implementation (and hence the reinforcement of this power position) and lack of knowledge of how to translate ideas into actions. It related to the lack of interest in the area by the leadership of the WaterAid Kiteto staff team, highlighting clearly how individual opinions and preferences are still able to influence practice within organizations and programmes.

Views on vulnerability and equity differ. Some think that when an agency works with a community in a participatory way this, by default, includes everyone in the community. Others understand communities to be differentiated, but disagree about whether it is the role of the international or local agency to work with the community to identify those who need subsidies or targeting in other ways because of their vulnerability or marginalization. Hence:

> Some view a gender-neutral participatory approach with pride, as non-intrusive and culturally sensitive. Yet in many cases, where participation has been pursued something is going wrong. Despite the stated intentions of social inclusion, it has become clear that many participatory development initiatives do not deal well with the complexity of community differences, including age, economic, religious, caste, ethnic and, in particular, gender. Looking back it is apparent that 'community' has often been viewed naively, or in practice dealt with, as a harmonious and internally equitable collective. Too often there has been an inadequate understanding of the internal dynamics and differences that are so crucial to positive outcomes. This mythical notion of community cohesion continues to permeate much participatory work, hiding a bias that favours the opinions and priorities of those with more power and the ability to voice themselves publicly. (Guijit and Shah 1999: 1)

But at the other end of the scale others believe that considering vulnerability should be core to all development work:

> In Kiteto, although trying to work our way through the myriad of complex relationships and sensitivities so as to be able to openly debate the pastoralist–agriculturalist issue, we were repeatedly pushed by a senior manager of one of our partners that we should be doing more in relation to understanding and responding to issues of vulnerability, particularly in relation to the difficulties faced by pastoralists and small peasant farmers in a district where land and water conflicts were becoming more commonplace.(House 1999 onwards, Kiteto, Tanzania)

Because of such wide-ranging views and no formal organizational policy committing programmes to consider gender, equity or vulnerability, whether these issues were considered in the various programmes seemed to depend mostly on the interest and commitment of individuals. In WaterAid's case, usually there was enough support in the line management chain to allow programmes to take the work as far as the staff and partners wanted. But where there was limited interest there was also limited pressure to ensure effective consideration.

The process of learning was also time-consuming and it was obvious that this process could have easily been derailed at points of staff changeover, or where the pressures of day-to-day work became too demanding. Limitations of time in the project cycle for water, hygiene and sanitation programmes could have been a constraint. However, the KINNAPA/KDC/WaterAid programme, was blessed with varying degrees of open-minded line management and organizational attitude that allowed appropriate time to be taken for building staff capacity, learning and responding to locally identified issues.

Organizational Change and Gender Activists

In terms of promoting gender within the organization and internationally, the debates faced similar barriers. International organizations probably find it more difficult to see and respond to gender and equity internally than within the project settings, where gender differences appear starker due to the often total absence of women. The difference is probably also related to the different power relationships involved when dealing with the internal workings of an international organization rather than project issues, where people want support in improving their water, hygiene or sanitation.

Gender activists also have their own limitations. As the gender debate often provokes heated discussion, there can be a lack of interest in becoming the champion or person who is seen to be pushing forward the debate. Becoming this person can often mean receiving the negativity that appears to be integral to the raising of this subject.

> Talking about gender inevitably reminds us at some level of our own – usually uneasy – position on the gendered power scale and the double binds involved in analysing that position and acting accordingly. We need to realise that extreme reactions of 'political correctness' or defensive dismissal of gender issues highlight the emotional risk to which people feel exposed when discussing power relations of which they are a part. (MacDonald et al. 1999: 49)

Without a person or team with organizational responsibilities to ensure the consideration of gender and equity throughout the organization, movement for change can become vulnerable or lost.

It also seems that those who are the most vocal in promoting gender equity are commonly the most strong headed, possibly those who have faced gender or equity issues themselves in the past. Although personal commitment is probably necessary to get the debate rolling in the face of sometimes quite strong resistance, it can also lead to differences of opinion about their work on gender between the activists themselves and there can then be a risk of self-implosion and even withdrawal from the debates. MacDonald et al. (1999: 35–54) identify the various types of change agents in the field of gender and provide interesting analysis and insights into gender dynamics within donor organizations themselves.

In international organizations often the senior and middle management are male and the more junior staff female. In WaterAid it was often female staff (although not always) who started debates about the need to consider gender and then later the more enlightened male staff joined as supporters and activists. However, due to the gendered divisions of power within organizations, if the female staff are unable to express their views and their male line managers are not in agreement, it is hard to progress with the gender issue.

It was obvious through the debates and processes in WaterAid in Tanzania that a lot of the interest and drive for pushing the gender debate forward was coming from the national female senior field staff and partner staff who were in positions just below management level. But to be able to move forward on the subject of gender, commitment and approval was also required from management level. In the case of WaterAid in Tanzania, the situation where a few management-level male staff disagreed with the issue of considering gender could have derailed the process which a national female senior field-level staff member had opened up through her efforts. Had there not also been other management team members keen to support her and to ensure that the debate was kept on the table her efforts may have been in vain. It was very interesting sitting in-between two national female senior field staff members as the debate re not considering gender was ensuing between a couple of the male management staff. Whispers indicating how much the female staff disagreed with the debate were numerous, but it was also obvious that some of the female staff members felt unable to challenge their line managers in this open forum. (House 1999, Dodoma, Tanzania)

Specific efforts to redress the balance of male and female staff in WaterAid Tanzania, at all levels, were made. These included ensuring there were always men and women on interview panels, providing additional funds for the employment of junior female field staff from marginalized ethnic groups and providing additional resources for capacity-building. These measures moved the situation forward but limitations were still found, including in recruitment opportunities for middle and senior management positions, with few women applicants applying for posts, limiting the selection pool.

Getting the debates started on gender and equity, getting commitment to the learning process and putting learning into action constitute a difficult road full of pitfalls. Reaching the point where staff were starting to respond to gender issues practically in the field, both within programmes and organizations, seemed from the field perspective quite an achievement in itself.

Understanding the Issues

Limited Guidance on How to Move Forward

Finding guidance on how to consider gender and equity in all aspects of programme activities on a daily basis is not that easy. On investigation advice can be found mainly for the initial analysis stages of power relations in publications, such as Slocum et al. 1995, and these are useful resources. But it is difficult to find guidance on how staff can integrate considerations of gender and equity into their everyday work or how to go beyond the initial analysis. What do you actually do when only men turn up to the meetings? How should you react when women are seen to be sitting all day to wait for water until the cattle have finished drinking? What should you do when the men or representatives of a majority ethnic group in a community take over decision-making and the voice of the minority or less powerful is dismissed? Gender analysis which limits itself to criticism that organizations working in water are not doing things correctly can help to keep gender on the agenda, which is in itself a positive contribution, but does not really help in taking the work in programmes and projects forward.

There are, however, some useful publications with practical examples of what can be done. These include the Asian and African field guides developed by the IRC (1994a and b), which provide clear case studies and examples of what can be done, and some upcoming publications edited by Reed (2005) and Reed and Coates (2005), which aim to make gender more accessible to technical personnel. Reed attempts to go one step further from theoretical analysis to practical responses and discusses the issues in plain language understandable by technical staff as well as social specialists. Other good ideas can be found elsewhere in individual or occasional papers (for example, Regmi and Fawcett 1999) and recently WaterAid pulled together some practical experiences into a gender manual for internal use (WaterAid 2004). However, the information currently available to help implementers to change their work practices remains generally difficult to access.

Understanding Context-specific Environments

To be able to effectively incorporate the gender and equity issue into program-mes and projects, the first step is for staff to look and learn in the communities and environments where they work. Undertaking basic field-level research can highlight the power and vulnerability issues, which will help to determine strategies to ensure that all groups are really heard and their views incorporated into the design, implementation and longer-term management of projects. This information collection could easily be incorporated into standard processes when first working with communities, but at present the way many tools such as participatory rural appraisal (PRA) are used only touches on such issues in a superficial way. Having women on the village PRA team may be taken as adequate evidence that women's views have been incorporated into the process, but significantly more effort is needed to help these women's views be heard and incorporated into the programme. This is also echoed by Guijit and Shah (1999).

In the village of Amei in Kiteto, a few women were included in the original village PRA team. However, it appears that no explicit effort was made to look at power relations or issues related to access, resources and decision-making and how the project should respond. After the project had progressed through to rehabilitation of the borehole system, whilst in the village for some finishing off of construction work, a WaterAid Engineer and other partner staff observed that the women were sitting for most of the day before they could get water. On questioning it was found out that the system developed, which involved men paying for diesel to pump water for their cattle and women being able to get water free, was not as positive a system for women as it seemed on first observation. Because the men were paying for the water for their cattle some were reluctant to allow women to take the water before their cattle had drunk. This was reported by women in Ndedo village to also be common practice at traditional wells in the district where women have to get out of the way and wait until the cattle have drunk, even if this means waiting for a large proportion of the day before they can also collect water. The fact that the team were not aware of this when they worked with the men and women of Amei to set up the system, or some were aware of it coming from the same cultural background, but did not think it was a relevant issue for the project, showed a serious weakness on the part of the project process. (House 2000, Kiteto, Tanzania, reported in various case studies in KINNAPA/KDC/WaterAid, 2002a)

The issue noted above for the pastoralist community of Amei is not the same issue, for example, as would be found in some of the agricultural-based communities who also live in Kiteto. Learning what the issues are takes time, good questioning and listening skills, and continual observation. Gender and equity are dynamic concepts and learning needs to start from the initial contacts with communities and continue through to the end of the project.

The Complexity of Gender and Equity Issues

Gender and equity are complex and intertwined with issues of ethnicity, age, culture, tradition, wealth and so forth. It was interesting working in Kiteto where there were stark differences in tradition/culture and gender between the three main types of communities, classified according to the basis of their livelihood: pastoralism, agriculture or hunter-gathering. Due to many factors such as pressure on land for farming, mobility, lack of take up of formal education leading to difficulties in engaging in national policy debates, and prejudice against non-farming ways of life leading to them being seen as inferior, the pastoralists, and even more the hunter-gatherer communities, had become marginalized. Conflicts were becoming more commonplace as land was changing from pastureland to farmland. However, whilst working with the pastoralists in particular, helping them to have more of a voice in discussions with the farming communities and their representatives, it was obvious that the concepts of marginalization and vulnerability were not clear cut. Within pastoralist society there were some very rich pastoralists and some very poor pastoralists, just as in farming communities there were rich and poor farmers. Rich pastoralists with large cattle herds wielded significant power in their communities and access and control issues over new projects were complicated by this power. Within both groups there were significant gender issues, with women from most ethnic groups in both pastoralist and farming communities having very little if any ownership or control of household resources, even of produce from land they may have farmed (Ngurumwa et al. 2001). Land taken over for large farms was taken not just by rich people from farming backgrounds, but also by rich people from pastoralist backgrounds.

Trying to ensure that the most vulnerable and marginalized had a voice in project activities in practice sometimes felt like holding a very delicate balance, trying to discuss issues related to marginalization and power, without further developing prejudices against or marginalization of any group.

Developing Methodologies and Skills and Commitment of Programme and Field Staff

Training

Working on gender and equity in water projects usually requires basic training for the staff so that they have some understanding of what the gender and equity issues are. They need to be able to see their work through 'gender aware eyes'. In Kiteto the route to training came through an acknowledgement that there were problems but that as a programme team we didn't know how to respond to them. A few of the team who were already reasonably gender aware undertook some field-based research in a number of villages. The findings of this research were then used as a starting point for the field team members to participate in awareness raising workshops.

It was obvious during the Kiteto programme and other country programme workshops (and even at international conferences) that there is a range of understanding on what gender is and a tendency for gender discussions to often cover the same ground. This was sometimes disillusioning. How many times were we going to discuss and agree that the important issue is that gender is about power relations between men and women and the way they interact and not just about women? While understanding basic concepts was important, moving on from the theories to practical ways of responding, and finding methodologies that work on the ground felt like significant steps.

Developing Methodologies

To be able to respond to the issues on the ground, the field team had to develop its own practical methodologies. These included:

- ensuring that, where there was evidence of exclusion, the teams communicated with all key groups in communities separately about project activities;
- supporting open discussions over difficult issues between representatives of minority and majority groups;
- postponing meetings where women were not present or in a severe minority, and discussing openly, and investigating with women, why they had not attended;
- openly congratulating/praising women in their ideas in open forums to build confidence;
- supporting representation of women as well as men in the more powerful committee roles and providing training;
- encouraging the men and women community representatives to openly monitor the participation of the various key groups in the community;
- including discussions on gender and equity aspects in all community trainings and continually raising and discussing issues in meetings;
- including male and female elders from the range of groups within the communities concerned in key decision-making processes particularly in relation to key sensitive issues (KINNAPA/KDC/WaterAid, Kiteto 2002a).

After a period of time the programme also documented the methodologies that had been identified, agreed which were being used and which were not, evaluated their effectiveness and decided which should be used in the future. The methodologies were discussed using verbal case studies to highlight where the different methodologies had been used and what had resulted. For wider organizational learning the outcomes of these discussions were documented and shared with other country programmes in WaterAid through an email discussion group.

Personal Values and Commitment

Learning about gender and equity is not the same as learning about a technical subject. Our views on the concepts and reasoning behind considering gender and equity are influenced by our own positions in society, our traditions and our cultures, in a way that the technical design of a pipeline usually is not. This means that basic understanding and commitment to the work will vary person to person. This is where it is clear that there is a particular value in facilitating and encouraging field staff to discuss their opinions and experiences openly between themselves. And the multiplier effect of, for example, a male pastoralist member of staff being a committed gender advocate, can be much more valuable than an international female staff member being the same. 'The encouragement of continued discussion on the gender issue, particularly between the field team members and with the open support of the partnerships management, seemed to have a significant effect on the teams' capacity and confidence' (House 2002, Kiteto, Tanzania).

An effort was made to keep ensuring that gender and equity were on the agenda. This was done through specific gender workshops allowing for reflection and sharing, or making sure that the subjects were continually considered during other activities such as meetings, planning events, general discussions, further research, and field work.

Interest in the subject and ownership of the work that the programme was undertaking on gender and equity took a while to get moving. However, after some time the fact that the team was trying hard to respond to gender and equity in the field started to become an area of a pride for the field team members and this helped to develop interest and confidence further.

One example of the sharing of experiences occurred through a competition set up for the field team members to document their experiences as case studies. Although most field staff had never written case studies before, most staff from partner organizations, KINNAPA and KDC, and WaterAid, energetically submitted them. The case studies were often enlightening and new ideas as well as problems were once again shared across the group, as shown in the following extracts:

When I was discussing this issue directly with women right at the meeting, so that one could stand up and give the answers to the questions I posed to them, one man told me that it is not possible for a woman to stand in front of a men's gathering. This is because, according to their beliefs and traditions, if that will happen, then all men at that meeting will die.

After that sentence I felt very uncomfortable and subordinated and the same to all women who were there. After that situation I started mobilizing the community about women's rights, their importance in development projects and donors views on gender aspects.

After that [on the third day] I went to the meeting to explain such a situation to the gathering and to make sure that we insisted on a sustainable way and the importance of women's participation and contribution towards the public meetings. This was whilst standing up and flushing out the men's minds about the wrong concept of dying while women stand up in front of them during meetings. I did this by asking them 'who died last night as the result of women standing in front of them yesterday?' (Paulina Ngurumwa in KINNAPA/KDC/WaterAid Kiteto 2002a)

As a facilitation team we asked the meeting to split into two groups of men and women separately. Our aim was to give more chance and freer opportunity to women to discuss and give suggestions and on top of that to make their own decisions on how to solve the problems. In the groups, female TUWI members facilitated the women's group and the male TUWI members went to the men's group. We spent almost one hour to facilitate discussions in the groups. Oh, it was very interesting to see how women were very active to talk in their group. And they made very strong decisions for improvement of the scheme management.

In fact from the decisions made by the women's group, when presented to the general meeting with men, they helped very much to prepare basic contents of the project management scheme. In the general meeting the team gave a chance to a women's representative to make feedback. She looked very confident. And in a very great extent, men in the meeting agreed with the decisions that were made by women. So instead of men seeing that the decisions were made by the individual woman who was presenting, they respected the decisions as a group decision. (Saad Makwali in KINNAPA/KDC/WaterAid Kiteto 2002a)

Developing Policies, Strategies and Commitment Statements

In WaterAid Tanzania the inclusion of a statement in the WaterAid Country Strategy happened after some time of working on gender across the programmes (the Kiteto programme was one of four) and investigating the issues. This was a significant step forward in the work of gender and vulnerability in the Country Programme.

WaterAid and its partners have already committed themselves to incorporating gender into all stages of the project cycle. We need to continue to develop strategies and monitoring tools that ensure the commitment is translated into real benefits for women and children. We also need to ensure that our commitment to sustainable development by means of empowering of the most vulnerable is constant, unwavering and solid. (WaterAid Tanzania 2001: 15)

In Kiteto, the partnership also developed its own gender and equity commitment statement, which was signed by the representatives of the three organizations. This commitment statement noted what staff believed in relation to gender and equity, the limitations of what they would be able to do, and what staff committed themselves to the best of their ability to do:

We believe that ... due to current power relations ... some groups participate easier than others, such as men, the more wealthy, and educated and that other groups, such as women, elderly, the very poor, the disabled and those in minority situations, will need additional support and encouragement to be able to participate.

But we also understand that ... the communities in which we work are very large [1,000–20,000 people] and that not every person will be able to participate in the day-to-day decision-making and activities in the project. But key decisions should be facilitated through the General Assemblies, which allow people some access to decision-making. However, broad representation in the project is a key step towards enhancing solidarity and the ownership across the social groups.

In response to these beliefs the programme teams commit, to the best of their ability, to try to ... ask questions when working in the villages and seeing that certain groups are

not being involved, as to why they are not and try to find out why. To find out whether
it is by their choice, by their priorities, or by exclusion, or lack of information, or other.
(KINNAPA/KDC/WaterAid Kiteto 2002a)

Although it was useful to have worked through the issues relating to gender
and equity as a team before coming up with the programme commitment statement, it was interesting to note the sigh of relief that one field worker made when
the commitment statement was signed by the programme's management. She
noted that now the team could feel confident in their work on gender and equity
in the villages, as they knew that they had the management's backing. This once
again highlighted the vulnerability of the consideration of gender and equity when
organizations have not engrained their commitment to these in their policies and
strategies.

As yet WaterAid still does not have an organizational policy on gender or equity
(see Chapter 7). Some developments in documenting experiences over the past
few years have been a start, but without leadership at senior management level
with a clear commitment to these issues, it is doubtful that institutional learning
will progress very far, or develop into organization-wide policies and strategies.

Successes and Weaknesses in Kiteto and the Communities' Views

In terms of ensuring that women and men from all livelihood backgrounds and
from across the social groups could participate in the project, some significant
progress was made over a period of three years.

> When I joined the team in the year 2001 and participated in the different activities, I
> realised that the Njoro community had changed a lot [Tuke had been involved in Njoro
> in 1994 and subsequent years through other KINNAPA project work]. It is also the only
> village of the programme with a pump attendant who is a woman, and also women
> who are members of the water committee meetings and in project activities they are
> now participating fully and contribute their ideas openly etc. So I can now conclude by
> saying that women in Njoro are recognised and respected and they have confidence too.
> (Gabrielle ole Tuke in KINNAPA/KDC/WaterAid Kiteto 2002a)

However, as with any process which involves a form of social transformation,
change does not happen overnight and strong views are still held on the various
sides in the communities where we worked. 'Some community members are
saying that to involve women is the way of under-grading men' (Hassan Mohamed
in KINNAPA/KDC/WaterAid Kiteto 2002a).

In 2002 a group of men and women in Njoro village were asked for their views
on aspects of the project and of the way the team worked on gender in the water

project. In this village in 1999, nearly all meetings were held with only men from the majority agricultural side of the community. By 2002, the pastoralist minority and the women were playing an increasing role and it was noted that sometimes the women even appeared stronger than the men. A selection of the range of views expressed follows (KINNAPA/KDC/WaterAid Kiteto 2002a):

> I am a wife, but today I am singing and dancing in front of people and I feel proud. [This woman led one of the choirs which sang and danced during the celebration.]

> The work that TUWI[3] the programme has been doing on gender has been very influential to our general lives as well – it has helped [in the relationships] in our homes. [Man]

> There are two levels of understanding in the village – that of the people here who are mainly the leadership who now have a good understanding of the issues and are now very committed, and then that of the general community where gender is still a problem. Even in the celebration today some men would not let their wives attend the celebration and they would have to remain at home. [Man]

> You should tell them that you should continue to facilitate women and men to be able to work together. I would like to give my personal experience. I am a Water Committee member and early on for a training session my husband would not allow me to attend the training. After he received education on gender he now allows me to go for training. My husband even cooked for my children when I went to a training. [Woman]

> There should be no separate work for men and women when distributing work. If someone says this to you [that women have too much work already and so they shouldn't be involved] then tell them that we said that you should continue working as you are with gender – and tell them not to take us backwards. [Woman]

The programme did not focus much on involving the young, elderly or disabled although this was changing by the end of 2002, particularly with the increased involvement of male and female elders in some of the communities. But the very poorest remain the hardest group to reach and pose a significant challenge, particularly when only limited numbers of people can participate in schemes in larger village. Methodologies still need to be further developed and refined to ensure that this group is reached.

Post 2002 – Results of Working on Gender and Equity

At the time of writing, it is too early to be able to determine the long-term impacts for the various communities of the way the teams worked on gender and equity. It is also difficult to determine the specific reasons why projects succeeded or failed, as there are so many complex aspects to communities and projects. However, the responses of the Njoro community noted above highlight some of the feelings of this particular community on the efforts of the team in the area of gender and equity.

Nakwetikya of Ndedo village also expressed her views in an interview for the WaterAid *Oasis* magazine, Spring/Summer 2003 (WaterAid 2003b):

> Three years ago, before we formed a committee and prepared ourselves as a community for the water source, men just saw women as animals. I think they thought of us as bats flapping around them. They had no respect for us and no-one would allow you to speak or listen to what you had to say. When I stand up now in a group meeting I am not an animal. I am a woman with a valid opinion. We have been encouraged and trained and the whole community has learnt to understand us.
>
> Oh, when I think back to how we used to feel terrible. I was treated like a donkey only fit to carry baggage all the time. Or a scrap of paper ... just rubbish in the wind. I can assure you though, that if you come back in a few years you will see that women will be the leaders in this village. That will bring so many benefits to everyone!

In early 2004 a number of the technical staff working on the programme were specifically asked whether they still considered gender and equity in their work. The respondents, all technicians from the Kiteto District Council, said they did (Makamba 2004). They gave examples from a number of projects, which they had worked on post 2002, where they had used principles learnt under the partnership and other programmes, to ensure the participation of men and women. On one occasion they had postponed a meeting in Namelock village because women were not present; in the following meeting a large number of women participated. In the two newer programme villages of the KINNAPA/KDC/WaterAid programme, Matui and Ndaleta, it was reported that two of the village pump attendants (paid workers) in both villages were women. In the previous four villages, before the teams had progressed very far with the gender work, the village pump attendants who were selected by the villages were all male, with one exception. The work of the KINNAPA/KDC/WaterAid programme in gender and equity, along with capacity-building from other programmes in the district, has definitely contributed to increasing the teams' capacity to be able to respond to gender issues in the field.

However, there have also been some less positive changes since mid 2002. Two young Maasai women were employed in KINNAPA Development Programme in the Gender, Women and Children's Affairs Unit and the Water, Hygiene and Sanitation Unit during 2000, in an attempt to build the capacity of the field teams, and to increase the representation of female and pastoralist staff on the programme, especially because in the pastoralist villages many Maasai women do not speak the national language, Kiswahili, and can be the most difficult to reach. Both of these staff were doing really well and contributing significantly to the programme's success. However, in 2003/2004, following a WaterAid decision to change the way it was going to fund its partners in Kiteto, the KINNAPA Board decided to make the two female staff redundant, along with another member of staff, and use the money saved to employ a new member of staff trained in livestock to continue the water and sanitation work. These decisions and changes highlight how vulnerable work on gender and equity is to changes in staff and situations, even when every effort has been made to ensure the reasoning behind the actions has been well understood and agreed, and agreements have been made in writing.

As a management-level staff member with capacity-building responsibilities and working with teams on the ground, when considering all of the barriers as a team we faced and responded to in Kiteto, simply to be able to respond reasonably effectively to gender and equity, I sometimes wonder how we managed as much as we did. As someone committed to considering gender and equity in technical projects, I have to admit that on occasions, with the massive workload at programme implementation level and all of the other barriers from within organizations and without, I wondered if perhaps this challenge would beat us. This concerns me as I wonder how implementation teams would manage if the leadership were not interested in, or so fully committed to, gender and equity, which sadly still appears in many cases to be the case. As a technical practitioner my plea goes at this point to the gender specialists to ask for a move, from the debates at theoretical levels and simple analyses and identifying that we're not doing it right, on to an approach which helps the practitioners to identify more practical ways of responding on a daily basis. This is really needed and without it I'm not sure how much progress will happen. (House 2004)

Conclusions

The barriers to considering gender and equity in projects and ensuring positive impacts are numerous. They include lack of clear policy, strategy and commitment at the organizational level and a lack of agreement as to whether gender should be considered at all. The lack of a person with specific responsibility to ensure this work is carried forward throughout the organization also causes limitations, as do the personal limits of gender activists themselves. There is also a lack of guidance available as to how to respond practically to gender and equity, making the gulf between the desires of the gender theorists and the capacities of the general practitioners to respond even wider. This can lead to long-standing conflicts between the two groups and contribute to non-action on the part of the practitioners.

There is a need to understand the context-specific environments to be able to respond appropriately in each scenario. This takes skill and experience and is not helped by the fact that gender and equity are complex subjects and are intertwined with power issues, which makes responding to them even more challenging. The process of building confidence, commitment and capacity of staff needs support over long periods of time. Gender and equity issues are linked to our own cultures and personal value systems and these make the learning and capacity-building processes even more complex, but also all the more important.

For all of the above barriers to be breached, the commitment of senior management is essential. However, even today after several decades of debate, this commitment is still not always there. It is important not to underestimate the difficulties often originating within institutions themselves with regards to attitudes towards and capacity for implementing programmes that are gender aware and that respond effectively to these issues. However, with continued efforts, commitment and resources, positive impacts can be made and appreciated by the women and men in the communities in which we work. Working on gender and equity – easier to say, harder to do, but still possible with significant effort.

Notes

1. An early version of this paper was presented at the Alternative Water Forum, 1 and 2 May 2003, at the Bradford Centre for International Development, University of Bradford, UK.
2. The views expressed in this chapter are those of the author and not necessarily of WaterAid as an organization.
3. TUWI is the KINNAPA/KDC/WaterAid Kiteto partnership's district community mobilization team made up of staff from all three organizations, KINNAPA Development Programme, Kiteto District Council and WaterAid.

References

Gender and Water Alliance (2003), 'The Gender and Water Development Report, 2003: Gender perspectives on policies in the water sector', WEDC for the Gender and Water Alliance.

Guijit, I. and M. K. Shah (eds) (1999), *The Myth of Community; Gender Issues in Participatory Development*, London: IT Publications.

House, S. (1999–2004), Personal observations, dated in text to reference the date of observation.

IRC (1994a), *Together for Water and Sanitation, Tools to Apply a Gender Approach,* Occasional Paper Series 24, The Asian Experience, IRC and Netwas.

—— (1994b), *Working with Women and Men on Water and Sanitation, An African Field Guide,* Occasional Paper Series 25, IRC and Netwas.

KINNAPA/KDC/WaterAid Kiteto (2002a), *Gender and Equity in the KINNAPA/KDC/ WaterAid Programme, Arusha, Tanzania,* includes case studies from implementing teams, methodologies tried and suggested with examples, and the programme and equity statement, collated by S. House.

—— (2002b) *Strategy Planning Workshop, 11–15th February 2002 for the KINNAPA/ KDC/WaterAid Partnership of Kiteto District, Arusha Region, Tanzania,* Report collated by E. Mmanda, S. House, D. Makamba, S. Ndeleya and K. Maganga.

MacDonald, M., E. Sprenger and I. Dubel (1999), *Gender and Organizational Change; Bridging the Gap between Policy and Practice,* second edition, The Netherlands: Royal Tropical Institute.

Makamba, D. (2004), Personal communication on interview points with the KINNAPA/ KDC/WaterAid Technical Support Team, Abdahlah Issa, Mr E.J. Kessy, Hassan Mohammed, Mihayo Sahani, and Mrs Mfinanga, and additional information on the progress of the programme and teams.

Ngurumwa, P.P., O.B. Mlekwa, G.M ole Tuke, M. Simon (2001), *Kiteto Gender Research (English Report), KINNAPA/KDC/WaterAid, 15–19/6/2000.*

Reed, B.J. (ed.) (2005), *Infrastructure for All: Meeting the Needs of Both Men and Women in Development Projects – A Practical Guide for Engineers, Technicians and Project Managers,* WEDC, Loughborough University, UK.

Reed, B.J. and S. Coates (2005), *Developing Engineers and Technicians: Notes on Giving Guidance to Engineers and Technicians on how Infrastructure can Meet the Needs of Both Men and Women,* WEDC, Loughborough University, UK.

Regmi, S.C. and B. Fawcett (1999), 'Integrating Gender Needs into Drinking Water Projects in Nepal', in *Gender and development,* 7(3): 62–72, Oxfam.

Slocum, R., L. Wichhart, D. Rocheleau, and B. Thomas-Slayter (eds) (1995), *Power, Process and Participation: Tools for Change,* London: IT Publications.

WaterAid (2003a), *Social Conflict and Water; Lessons from North-East Tanzania,* Discussion Paper.

—— (2003b), 'Out from the depths', *Oasis, The WaterAid Journal,* Spring/Summer 2003.

—— (2004), Gender Resource Pack, 2004. Unpublished.

WaterAid Tanzania (2001), *Country Strategy Paper,* strategy co-ordinated by Dave Mather.

Index

Note: *italicized* page numbers indicate tables